T0329819

# CHROMATOGRAPHY

# CHROMATOGRAPHY

Principles and Instrumentation

**Mark F. Vitha**

Published by John Wiley & Sons, Inc., Hoboken, New Jersey
Published simultaneously in Canada

For general information on our other products and services or for technical support, please contact our Customer Care Department within the United States at (800) 762-2974, outside the United States at (317) 572-3993 or fax (317) 572-4002.

Wiley also publishes its books in a variety of electronic formats. Some content that appears in print may not be available in electronic formats. For more information about Wiley products, visit our web site at www.wiley.com.

*Library of Congress Cataloging-in-Publication Data:*

Names: Vitha, Mark F.
Title: Chromatography : principles and instrumentation / Mark F. Vitha.
Description: Hoboken, New Jersey : John Wiley & Sons, Inc., [2017] | Includes
     bibliographical references and index.
Identifiers: LCCN 2016011576 | ISBN 9781119270881 (cloth : alk. paper) | ISBN
     9781119270898 (pdf : alk. paper) | ISBN 9781119270904 (epub : alk. paper)
Subjects: LCSH: Chromatographic analysis. | Gas chromatography. | Liquid
     chromatography.
Classification: LCC QD79.C4 V58 2017 | DDC 543/.8–dc23 LC record available at
     http://lccn.loc.gov/2016011576

About the cover: A two-dimensional liquid chromatography (LC × LC) separation of maize seed extract. Ground seed was extracted using methanol mixed with a phosphate buffer (pH 5.7) followed by solid phase extraction prior to analysis. The first dimension separation used a Zorbax SB-C3 column while a carbon-clad core-shell silica column was used for the second dimension. Additional experimental details about the extraction and chromatographic system can be found in Huang, Y.; Gu, H.; Filgueira, M.; Carr, P.W. *J. Chromatogr. A*, 1218, *2011*, 2984–2995 and Filgueira, M.; Huang, Y.; Witt K.; Castells C.; Carr, P.W. *Anal Chem.* 83, *2011*, 9531–9539. The author thanks the creators of this image for their permission to use this image and Marcelo Filgueira for the work he put into it for this book.

Typeset in 10/13pt PalatinoLTStd-Roman by SPi Global, Chennai, India

Printed in the United States of America

10 9 8 7 6 5 4 3 2 1

# CONTENTS

# PREFACE

Chromatography is the most widely used technique in modern analytical chemistry. In this book, you will learn the fundamental principles underpinning chromatography and be able to connect those principles to the design and use of chromatographic systems.

The focus of this book is current theory and the modern practice of chromatography. The successful practice of chromatography depends upon an understanding of molecular-level processes such as partitioning, band broadening, and the effects of temperature and mobile phase modifiers on solute retention. In every chapter, you will find figures and problems designed to help you visualize these processes and identify the key variables in any given chromatography problem.

Sections such as those on microfabricated gas chromatography (GC) separations, two-dimensional liquid chromatography (2D-LC), superficially porous particles, ultra-high performance liquid chromatography (UHPLC), and Orbitrap mass spectrometry ensure that you are introduced to the most recent developments in chromatography and are thus well prepared to enter the workforce.

Case studies illustrate the range of applications addressed by chromatography. Examples such as the analysis of performance-enhancing drugs in sports and the detection of deliberate contamination of food with melamine and Sudan dyes are discussed in the context of liquid chromatography (LC). Chapter 3 ends with a detailed discussion of the analysis of pharmaceutical compounds found in waterways across the United States as part of a U.S. Geological Survey study. Chapter 2 describes a case study centered on the analysis of Middle Eastern oil samples. This case was part of an international incident that involved the United States, Iraq, Iran, and Russia, and it shows the impact of analytical chemistry in general, and chromatography, specifically, on world affairs. All of the case studies demonstrate how the principles and instrumentation presented earlier in the chapters contribute to solving practical, real-world problems.

This book also complements a modular approach to teaching. Rather than being encyclopedic, this focused book allows instructors to integrate selected sections with other modular books focused on techniques such as spectroscopy or mass spectrometry, or to supplement the information provided in this book with their own materials. It also facilitates its use in courses such as interdisciplinary laboratory courses, biochemistry, and introductory analytical chemistry courses.

The book has three chapters. Chapter 1 provides an in-depth description of the processes governing chromatography. It includes a systematic development of band broadening, drawing extensively from the work of J. Calvin Giddings. Numerous figures help you visualize the processes that contribute to band broadening and link those processes to key equations. Problems embedded within the chapter reinforce the connections between theory, visualization, and the contributions of different factors to the overall breadth of a peak.

Chapter 2 focuses on gas chromatography. It introduces the components of GC instruments in the context of the function they accomplish and discusses the theory behind instrumentation, instrument design, and practical aspects such as temperature programming, injection volumes, and the selection of stationary phases and detectors that practitioners must consider when conducting GC analyses.

The final chapter addresses liquid chromatography. It emphasizes reversed-phase liquid chromatography (RPLC) because of its prominence, but you will also find descriptions of all of the common modes of LC, including hydrophilic interaction chromatography, which has great attributes for the separation of small, polar molecules, and which may prove to be quite valuable as a mode that is orthogonal to RPLC for two-dimensional separations. This chapter also discusses topics such as sub-2-μm particles, superficially porous particles, UHPLC, and 2D-LC as part of the mainstream modern practice of LC, preparing you with the knowledge needed to operate in modern laboratories.

There are many people to thank for the completion of this book. The sacrifices my mother made and her unflagging commitment to my education provided the foundation upon which this book exists. Each page of this book embodies her love, support, and dedication. My wife, Maura Lyons, has been a stalwart supporter throughout the process. Her love, encouragement, and even keel are irreplaceable. Thanks also to Greg Febbraro, who tracked the progress of this book through all its ups and downs. I wish we could celebrate its publication together – your friendship is missed. Frank Settle, whose *Instrumental Methods of Analysis* textbook provided the genesis for this project, has been unwavering in his support. Some figures and text from that work have been included in this book. I owe much to Peter Carr, both for the specific work he put into reviewing material in this book and for the years he has spent teaching me chromatography. Leah Carr fostered an academic family and I, like many, miss her support and affection. I am grateful to Joseph Brom and Gary Mabbott for the research opportunities they provided that introduced me to instrumentation generally and to separation science specifically. The influence of their teaching made this book possible. John Dorsey, Stephen Weber, Dwight Stoll, Charles Lucy, Paige Diamond, Teresa Golden, Brian Gregory, Brian Lamp, Yinfa Ma, David McCurdy, and James Miller all provided valuable feedback as reviewers. I appreciate the time and effort they spent on my behalf. Much of what is right in this book is thanks to them, and all errors are mine. Susan Boyer provided significant feedback regarding the prose. Dwight Stoll and Paul Boswell shared an Excel chromatogram generator that I used to create many of the chromatograms in this book. Neal Byington at the U.S. Customs and Border Protection was exceptionally helpful in discussions regarding the analysis of oil samples from the Middle East related to the case study in Chapter 2.

Mark F. Vitha
*Des Moines, Iowa*
*October 4, 2016*

# FUNDAMENTALS OF CHROMATOGRAPHY

**1**

Many "real-world" samples are mixtures of dozens, hundreds, or thousands of chemicals. For example, medication, gasoline, blood, cosmetics, and food products are all complex mixtures. Common analyses of such samples include quantifying the levels of drugs – both legal and illegal – in blood, identifying the components of gasoline as part of an arson investigation, and measuring pesticide levels in food.

Chromatography is a technique that separates the individual components in a complex mixture. Fundamental intermolecular interactions such as dispersion, hydrogen bonding, and dipole–dipole forces govern the separations. Once separated, the solutes can also be identified and quantified. Because of its ability to separate, quantify, and identify components, chromatography is one of the most important instrumental methods of analysis, both in terms of the number of instruments worldwide and the number of analyses conducted every day.

## 1.1. THEORY

Chromatography separates components in a sample by introducing a small volume of the sample at the start, or head, of a column. A mobile phase, either gas or liquid, is also introduced at the head of the column. When the mobile phase is a gas, the technique is referred to as gas chromatography (GC) and when it is a liquid, the technique is called liquid chromatography (LC). Unlike the sample, which is injected as a discrete volume, the mobile phase flows continuously through the column. It serves to push the molecules in the sample through the column so that they emerge, or "elute" from the other end.

Two particular modes of LC and GC, known as reversed-phase liquid chromatography (RPLC) and capillary gas chromatography, account for approximately 85% of all chromatographic analyses performed each day. Therefore, we focus on these two techniques here and leave discussions of specific variations to the chapters that describe LC and GC in greater detail.

In GC, the mobile phase, which is typically He, $N_2$, or $H_2$ gas, is delivered from a high-pressure gas tank. The gas flows through the column toward the low-pressure end. The column contains a stationary phase. In capillary GC, the stationary phase is typically

*Chromatography: Principles and Instrumentation*, First Edition. Mark F. Vitha.
© 2017 John Wiley & Sons, Inc. Published 2017 by John Wiley & Sons, Inc.

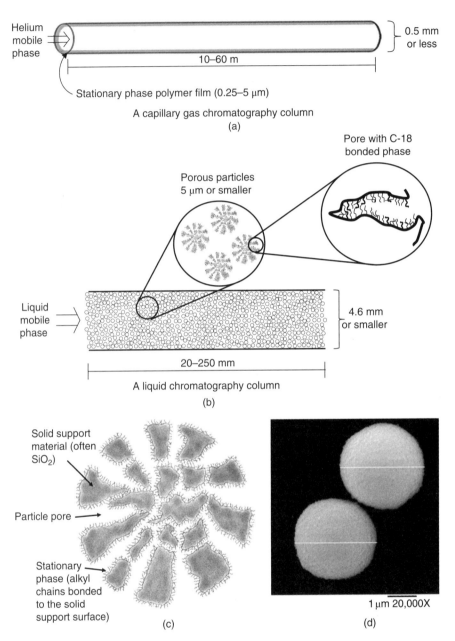

**FIGURE 1.1**   Representations of typical capillary gas (a) and liquid (b) chromatography columns. Figure (c) is a depiction of a cross section of a porous particle (shaded areas represent the solid support particles, white areas are the pores, and the squiggles on the surface are bonded alkyl chains. Figure (d) is an scanning electron microscope (SEM) image of actual 3 μm liquid chromatography porous particles. Note that the lines across the particle diameters have been added to the image and are not actually part of particles. (*Source:* Alon McCormick and Peter Carr. Reproduced with permission of U of MN.). It is worth taking time to note the different dimensions involved. For the GC columns, they range from microns ($10^{-6}$ m) for the thickness of the stationary phase, to millimeters ($10^{-3}$ m) for the column diameter, up to tens of meters for the column length. Note also that LC columns are typically much shorter than GC columns (centimeter versus meter).

a polymer film that is 0.25–5 μm thick (see Figure 1.1a). It is coated on the interior walls of a fused silica capillary column with an inner diameter of approximately 0.5 mm or smaller. The column is usually 10–60 m (30–180 ft) long.

RPLC is the most common mode of liquid chromatography. In RPLC, the mobile phase is a solvent mixture such as water with acetonitrile ($CH_3CN$) that is forced through the column using high-pressure pumps. The column is typically made of stainless steel, has an inner diameter of 4.6 mm or smaller, and is only 20–250 mm (1–10 in.) in length (see Figure 1.1b). However, unlike most GC columns, most LC columns are packed with tiny spherical particles approximately 5 μm in diameter or smaller, as shown in Figure 1.1c and d. When rubbed between your fingers, the particles feel like talc or other fine powders. The particles are not completely solid, but rather are highly porous, with thousands of pores in each particle. The pores create cavities akin to caves within the particle. The pores create a large amount of surface area inside the particles. A stationary phase, typically an alkyl chain 18 carbon atoms long, is bonded to the surface of these pores. A more specific discussion of the important aspects of these particles, and variations in the kinds of stationary phases bonded to them, is provided in Chapter 3. For now, it is simply important to have an image of a stainless steel column packed with very fine porous particles that have an organic-like layer bonded to the surface of the pores.

Some of the important RPLC and capillary GC column characteristics are summarized in Table 1.1. We also point out here that a chromatographic analysis is conducted with an instrument called a *chromatograph* and results in a *chromatogram*, which is a plot of the detector's response versus time (see Figure 1.2). Subsequent sections describe how retention and separation of molecules are quantified.

### 1.1.1. Component Separation

Different types of molecules are separated within the column because they have different strengths of intermolecular interactions with the mobile and stationary phases. To help

**TABLE 1.1  Common RPLC and GC Characteristics**

|  | RPLC | GC (open tubular) |
|---|---|---|
| Column construction | Stainless steel | Quartz with a polyimide coating |
| Column length | 20–250 mm | 10–60 m |
| Column inner diameter | 2.1– 4.6 mm | 0.1–0.5 mm |
| Particle composition | Porous silica ($SiO_2$) particles | No particles – open tube |
| Particle size | 1.8–5 μm | No particles – open tube |
| Mobile phase | Solvent mixture (e.g., water mixed with acetonitrile) | He, $N_2$, or $H_2$ |
| Stationary phase location | Alkyl chains (C-8 and C-18) bonded to particle surface | Liquid-like polymer film bonded to capillary walls |
| Stationary phase chemistry | Relatively nonpolar and organic in nature | Polysiloxane polymer derivatized with organic moieties |

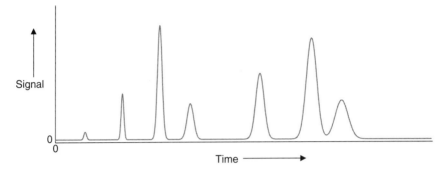

**FIGURE 1.2**   An example of a chromato*gram* – a plot of signal versus time – measured using a chromatograph (the instrument). Each peak represents a different solute that emerges from the column at a different time than the others. The peak width and height are related to the amount of each solute present.

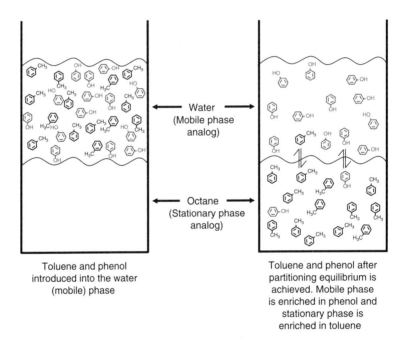

| Toluene and phenol introduced into the water (mobile) phase | Toluene and phenol after partitioning equilibrium is achieved. Mobile phase is enriched in phenol and stationary phase is enriched in toluene |

**FIGURE 1.3**   This figure depicts the behavior of phenol and toluene (solutes) partitioning between water and octane (bulk solvents). The water and octane serve as models for the mobile and stationary phases, respectively, in liquid chromatography. The left image depicts the system right after solutes are added to the aqueous phase before equilibrium is established. Once equilibrium is established (right), more toluene than phenol partitions into the nonpolar octane phase. Similarly, more phenol resides in the water due to hydrogen bonding and dipole-dipole interactions.

understand chromatographic separations, we first use a simplified model of liquid chromatography with water as the mobile phase and octane ($C_8H_{18}$) as the stationary phase. Imagine that a mixture of toluene and phenol is introduced as solutes into the mobile phase as depicted in Figure 1.3.

In this static image, given enough time, the solute molecules diffuse through the water and into the octane. They eventually reach equilibrium, being distributed to different extents between the water (mobile) and octane (stationary) phases. This equilibrium process is described in Equation 1.1

$$A_{mobile} \rightleftharpoons A_{stationary} \tag{1.1}$$

with the associated equilibrium constant

$$K = \frac{[A]_s}{[A]_m} \tag{1.2}$$

where "A" represents a specific analyte such as phenol or toluene, and $K$, by IUPAC definition, is known as the distribution constant. Many chromatographers refer to it as the partition coefficient or distribution coefficient. We will treat all of these as synonymous in this and the following chapters.

Because phenol is more polar than toluene and capable of hydrogen bonding with water, it does not partition into the octane to the extent that the toluene does. When looked at from a temporal perspective, phenol molecules spend less time in the octane, on average, than do the toluene molecules, which are attracted to the octane by dispersion interactions. It is important to understand that phenol is also attracted to the octane by dispersion interactions, and in fact, toluene is attracted to water through dispersion and dipole-induced dipole interactions. However, because phenol can participate in dipole–dipole and hydrogen-bonding interactions with water, and toluene cannot, phenol has a greater affinity for the aqueous phase than does toluene. As a consequence, phenol stays in the water more and partitions less into the stationary phase than does toluene.

It is clear from Figure 1.3 that what was once a mixture of an equal number of phenol and toluene molecules separates by differential partitioning between the mobile and stationary phases. It is easy to imagine that if the water phase were now drawn off and allowed to equilibrate with a fresh volume of octane, further purification of the phenol from the toluene would occur. Done repeatedly, eventually the phenol and toluene would be completely separated from one another.

In the actual practice of chromatography, these individual, discrete steps such as just described are not actually performed, but the effect of partitioning within a column is the same. As the mobile phase is continuously introduced into the column, the solutes continuously partition between it and the stationary phase. Because the molecules do not move down the column when they are in the stationary phase, those with higher affinities for the stationary phase relative to the mobile phase, meaning those with high distribution constants (i.e., partition coefficients), lag behind those with smaller distribution constants. In other words, some molecules elute from the column relatively quickly because their affinity for the mobile phase is greater than that for the stationary phase. Others, whose partitioning favors the stationary phase, take more time to make it through the column. For the molecules considered in Figure 1.3, phenol elutes before toluene. In this way, different molecules are separated within the column based on their intermolecular interactions with the stationary and mobile phases, which ultimately depends on the structure of the solutes and the chemical composition of the phases. The separated molecules are detected at the end of the column using a variety of detectors that can quantify, and in some cases, identify them as they elute. The detectors used in GC and LC are described in detail in the following chapters.

It is important to note that all molecules spend the same amount of time in the mobile phase. Therefore, *the separation occurs because of different times spent in the stationary phase.* To understand this, think about two canoes that start down a river at the same time, being carried along solely by the current. The time it takes them to go from one end of the river to the other, without stopping, is dictated simply by the speed of the current and the length of the river (distance = velocity × time). If along the way, the canoes pull into a number of ports (i.e., the stationary phase), they stop their progress down the river. If one canoe stops more than another, and stays in port for longer periods of time (i.e., has stronger interactions with the ports), then it reaches the end of the river later than the canoe that did not stop as often or for as long. Thus, a separation has occurred because the *total* time for the journey is different for the two canoes. But still, the time spent *on the river* (i.e., in the mobile phase) is the same because both canoes covered the same distance and moved at the same velocity when on the river. Therefore, the difference in the canoes' affinities for being in port (i.e., in the stationary phase) caused the separation.

It should be noted that the simplified mobile and stationary phases depicted in Figure 1.3 are rough approximations to actual liquid chromatography systems, which are discussed in more detail in Chapter 3. In addition, if the water is replaced with a gas such as helium or nitrogen, the system approximates conditions in gas chromatography. Thus, separations in both GC and LC depend on the relative strength of intermolecular interactions of solutes with the mobile and stationary phases. The only difference is that in GC, because molecules do not interact with the gaseous mobile phase, the separation is dictated solely by the relative strengths of interactions of solutes with the stationary phase.

### 1.1.2.  Retention Factor

In the preceding section, we established that different molecules spend different amounts of time traveling through the column. The total amount of time that a molecule spends in the column, from the time of injection to the time of detection, is called the retention time, $t_r$. The name indicates that we think of the molecules as being retained by the column – specifically by the stationary phase into which the solutes partition. In Figure 1.4, the retention time of phenol is 3.3 min and toluene's retention time is 5.2 min. While in the column, molecules spend their time in two places – the mobile phase and the stationary phase. Thus, the total time they are retained is simply the sum of the time they spend in each. Hence,

$$t_r = t_s + t_m \tag{1.3}$$

where $t_s$ and $t_m$ are the time spent in the stationary and mobile phases, respectively.

While the retention time is the most fundamental quantity measured, we often convert it into a dimensionless quantity called the retention factor, $k$, where

$$k = \frac{t_r - t_m}{t_m} \tag{1.4}$$

Here, $t_m$ is a measure of the time it takes the mobile phase to flow from the start of the column to the end of the column. It is often referred to as the "dead time," or "hold-up time." A solute that has no affinity for the stationary phase and therefore travels down the column at the same rate as the mobile phase is used to measure the dead time. This peak is marked as $t_m$ in Figure 1.4.

Converting retention times into retention factors normalizes for some operating conditions that vary between columns. For example, longer columns produce longer retention times even if everything else such as particle size, column diameter, and mobile phase flow rate are the same. More specifically, suppose one laboratory uses a column that is twice as long as the one used by another laboratory. In this case, $t_r$ and $t_m$ double because the molecules have twice the distance to travel. However, Equation 1.4 shows that $k$ is the same in both laboratories because

$$k = \frac{2t_r - 2t_m}{2t_m} \text{ (lab with longer column)}$$

$$= \frac{2(t_r - t_m)}{2t_m} = \frac{t_r - t_m}{t_m} \text{ (lab with shorter column)} \tag{1.5}$$

Another reason for focusing on retention factors rather than retention times is that $k$ is directly related to the distribution constant, which, as described above, fundamentally controls the separation process. To derive the relationship between the retention factor, $k$, and the distribution constant, $K$, note that Equation 1.3 can be rewritten as shown in Equation 1.6,

$$t_r - t_m = t_s \tag{1.6}$$

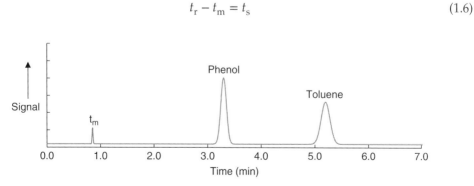

**FIGURE 1.4** Chromatogram of phenol and toluene. The retention times of phenol and toluene are 3.30 and 5.20 min, respectively. The dead time, the time it takes an unretained solute to pass through the column, is labeled as $t_m$.

so that Equation 1.4 can be rewritten as

$$k = \frac{t_s}{t_m} \qquad (1.7)$$

such that $k$ reflects the ratio of the time one type of molecule (e.g., phenol) spends in the stationary phase relative to the time it spends in the mobile phase. The longer the molecule spends in the stationary phase relative to the time it spends in the mobile phase, the greater the value of $k$.

For a collection of identical molecules – for example, 100 phenol molecules – the average time spent in the stationary phase relative to the mobile phase also reflects the instantaneous distribution of molecules between the phases. If molecules spend more time, on average, in the stationary phase than in the mobile phase, then a snapshot taken at a discrete point in time shows more molecules in the stationary phase ($n_s$) than in the mobile phase ($n_m$). Thus,

$$k = \frac{t_s}{t_m} = \frac{n_s}{n_m} \qquad (1.8)$$

Because the molar concentration of a solute, A, is given by $n_i/V_i$ where $n$ is the number of moles of A in phase i, $V$ is the volume of phase i, and "i" is either the mobile or stationary phase, combining Equations 1.2 and 1.8 results in

$$k = \frac{\left[\dfrac{n_s}{V_s}\right]V_s}{\left[\dfrac{n_m}{V_m}\right]V_m} = \frac{[A]_s V_s}{[A]_m V_m} = K\left(\frac{V_s}{V_m}\right) = K/\beta \qquad (1.9)$$

where $\beta = V_m/V_s$ and is called the "phase ratio." It is important to note that this definition of the phase ratio is the one given by the International Union of Pure and Applied Chemistry (IUPAC), but because many people define the phase ratio as $\phi = V_s/V_m$ such that $k = K\phi$, it is important to understand which definition is being used in different publications. In this and subsequent chapters, we will follow the IUPAC definition.

The phase ratio plays a role in retention that can be understood using chemical reasoning and logic. Going back to Figure 1.3, if more octane is added to the beaker, the ratio of octane to water increases. Furthermore, just out of sheer probability, toluene and phenol molecules respond to the addition of octane by partitioning out of the water into the octane. Taken to the extreme, as the mobile phase volume goes to zero, all of the molecules *have to* partition into the stationary phase, with the consequence that $t_s$ increases and $t_m$ decreases as shown in Figure 1.5. The result is that $t_r$ increases for all solutes. Similarly, as $V_s$ decreases relative to $V_m$, the distribution of solutes shifts toward the mobile phase and $t_r$ decreases.

In practice, the physical characteristics of the column that one purchases dictate the phase ratio and cannot be manipulated easily. To change the phase ratio, a different column must be purchased and installed in the instrument. Manipulating retention in this manner is more frequently a consideration in gas chromatography than it is in liquid chromatography. In GC, stationary phases with thicker polymer films bonded to the capillary

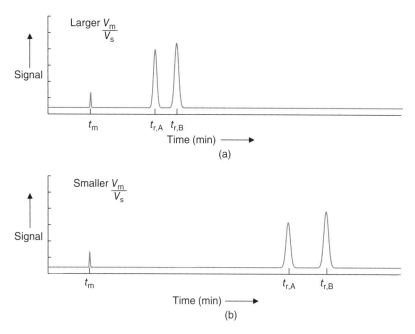

**FIGURE 1.5** Effect of decreasing the phase ratio ($\beta$, $V_m/V_s$) by increasing $V_s$ and decreasing $V_m$ on retention (assuming constant $V_{tot}$). Solute retention times, $t_r$, increase as the volume of stationary phase ($V_s$) in the column increases due to solutes spending more time in the stationary phase.

walls that increase retention are available. In LC, most columns have bonded phases that are 18 carbon atoms long, although shorter chains that produce less retention are available.

## EXAMPLE 1.1

We have used water and octane as models of a mobile and stationary phase in the text and we use it again in this question. Suppose 10,000 molecules each of butylamine and methyl hexanoate are added to a closed system containing 500 mL each of water and octane. The system is then shaken and allowed to come to equilibrium. The water-to-octane distribution constant for butylamine is 0.260 and 171.8 for methyl hexanoate.[1]

(a) How many molecules of butylamine and methyl hexanoate are in the aqueous phase at equilibrium?
(b) Based on the calculations, which solute is more concentrated in the aqueous phase at equilibrium?
(c) Which in the stationary phase?
(d) Based on the structures of the molecules and using arguments regarding their intermolecular interactions, rationalize the differences in the distribution constants for the two molecules.
(e) If more octane than water was used in the system, would the number of moles of each solute in the water phase increase or decrease?

**Answer:**

(a)

$$K = \frac{[\text{butylamine}]_{\text{octane}}}{[\text{butylamine}]_{\text{water}}} = \frac{\left(\frac{n_{\text{octane}}}{V_{\text{octane}}}\right)}{\left(\frac{n_{\text{water}}}{V_{\text{water}}}\right)} = 0.260$$

$$n_{\text{octane}} + n_{\text{water}} = 10{,}000 \text{ so } n_{\text{octane}} = 10{,}000 - n_{\text{water}}$$

And we know $V_{\text{octane}} = V_{\text{water}}$ so the volumes cancel. Substituting yields:

$$K = \frac{10{,}000 - n_{\text{water}}}{n_{\text{water}}} = 0.260$$

$$10{,}000 = 0.260 n_{\text{water}} + n_{\text{water}} = 1.26 n_{\text{water}}$$

$7940 = n_{\text{water}}$ for butylamine (approximately after rounding to three

significant figures)

Following the same procedure yields ~58 molecules of methyl hexanoate in the aqueous phase at equilibrium, meaning the other 9942 methyl hexanoate molecules are in the octane.

(b) Butylamine is much more concentrated in the aqueous (mobile) phase, while methyl hexanoate is much more concentrated in the octane/stationary phase.

(c) Octane is being used to represent a stationary phase.

(d) Butylamine is polar, can donate hydrogen bonds to water, can accept hydrogen bonds from water, and has fewer carbon atoms, so it can interact well with water. Methyl hexanoate is a larger compound with more carbon atoms, giving it stronger dispersion interactions with the octane. It is also polar (although less so than butylamine) and can hydrogen bond, but these effects are not as strong and therefore water does not compete as well as the octane to attract the methyl hexanoate.

(e) The phase ratio would shift in favor of the octane, so more molecules of both solutes would be found in the octane phase at equilibrium, meaning fewer than what was calculated in part (a) would be in the aqueous phase.

**Another question:**

The retention time of a solute is measured to be 23.76 min using gas chromatography. Under the same conditions, the dead time is 0.88 min.

(a) What is the ratio of the moles of solute in the stationary phase relative to the moles of solute in the mobile phase at any point in time during the analysis?

(b) If the column were changed to one with a larger phase ratio (i.e., larger $V_m/V_s$), would this ratio increase or decrease?

(c) Would the retention time increase or decrease?

**Answer:**

(a) $k = \frac{n_s}{n_m} = 26.0$ (or 26 : 1).

(b) Equation 1.9 shows that as $\beta$ increases, $k$ decreases, so the ratio of solutes in the stationary phase relative to those in the mobile phase also decreases. An increasing phase ratio means greater volume of mobile phase relative to stationary phase. Therefore, decreasing retention makes sense because there is less stationary phase present to retain the molecules and/or more mobile phase present for solutes to partition into.

(c) As a consequence, the retention time of the solute decreases.

### 1.1.3. Separation

The *retention* of any single component is typically not of primary importance to a chromatographer. The idea of *separation* is much more important. In other words, do different types of molecules elute at sufficiently different times so that they can be individually quantified and identified?

As Equations 1.4–1.9 show, molecules that have higher distribution constants (larger $K$)

1. have higher retention factors (larger $k$), and hence
2. are retained longer, causing them to
3. have higher retention times, $t_r$,

than those with smaller distribution constants. In Figure 1.4, toluene has a higher distribution constant as established in Figure 1.3 and thus elutes later than phenol. Molecules that are retained for a long time are likely to be well separated from those that are retained for a short time.

The degree of separation between any two solutes, A and B, is quantified using a parameter called the separation factor, $\alpha$.

$$\alpha = \frac{k_B}{k_A} \tag{1.10}$$

where "B" is the solute with the longer retention time and higher $k$. The "separation factor" is also frequently called "selectivity," although IUPAC discourages this use. The words "separation factor" and "selectivity" are quite descriptive and convey the idea that the ratio measures the extent of separation between two solutes – in other words, how selectively one compound is retained relative to another on the same column.

Molecules that are well separated from one another have high separation factors, while solutes that elute close to each other have small separation factors. However, the difference in retention times is not the only factor that has to be considered when measuring separation.

Consider the two chromatograms in Figure 1.6. The retention factors of both solutes are the same in the two different chromatograms. Because the retention factors are the same, $\alpha$ is the same. Yet, the separation in Figure 1.6b is incomplete – baseline separation has not been achieved – but in Figure 1.6a the solutes are fully resolved. Clearly, the separation factor is not the only parameter that dictates how well resolved one component is from another.

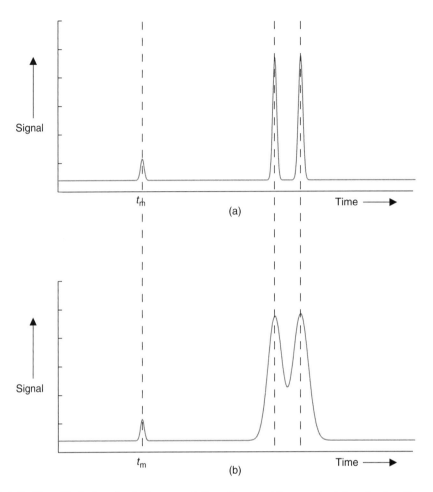

**FIGURE 1.6**  The effect of peak width on resolution. The dead time and solute retention times are the same in both chromatograms, meaning that the separation factor is the same in both chromatograms. The lack of resolution in (b) compared to (a) is therefore due to the widths of the peaks.

## EXAMPLE *1.2*

(a) Estimate the retention factors of phenol and toluene in Figure 1.4.
(b) What is the separation factor of the separation?

**Answer:**

(a)
$$k_{phenol} = \frac{t_r - t_m}{t_m} = \frac{3.30\,min - 0.90\,min}{0.90\,min} = 2.7$$
$$k_{toluene} = \frac{t_r - t_m}{t_m} = \frac{5.20\,min - 0.90\,min}{0.90\,min} = 4.8$$

(b) $\alpha = \dfrac{4.8}{2.7} = 1.8.$

**Another question:**

(a) What is the new value for the separation factor if the separation is repeated under conditions that lead to $k = 7.73$ and $k = 9.14$ for phenol and toluene, respectively?

(b) Which separation has better (i.e., higher) selectivity?

**Answer:**

(a) $\alpha = 1.18$.

(b) The first separation has better selectivity.

### 1.1.4. Resolution and Theoretical Plates

Resolution between peaks is important because it makes it possible to quantify each individual component in a mixture more accurately and precisely than when peaks overlap. In the extreme case of complete overlap, quantitation is typically completely impossible (except when a mass spectrometer or an analyte-selective detector is used), and the scientist might not even be aware that the observed peak is the result of multiple components.

Figure 1.6b makes it clear that in addition to the separation factor, the width of the peaks is important in terms of resolution. Thus, a consideration of the processes contributing to peak widths follows.

In order to fully understand the physical processes that occur within a chromatographic column that lead to peak broadening, also known as "band broadening," it is first important to be able to quantify peak widths. Ideally, chromatographic peaks elute with a Gaussian profile, as shown in Figure 1.7. Two measures of peak widths are commonly

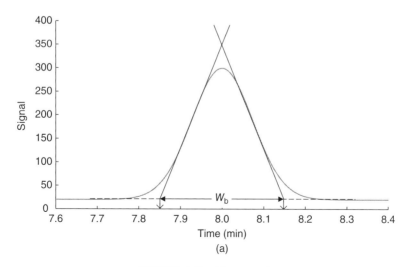

**FIGURE 1.7** *(Continued)*

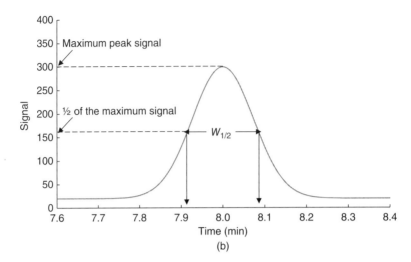

**FIGURE 1.7** Illustration of the two different measurements of width: (a) baseline width ($W_b$) and (b) full width at half maximum (FWHM, $W_{1/2}$).

used, the baseline peak width ($W_b$) and the peak width at half maximum height ($W_{1/2}$). The baseline peak width is found by drawing tangents to the curve as shown in Figure 1.7a and determining the distance between the two tangents at the baseline of the peak. To find $W_{1/2}$ the highest signal reached by the peak is found (300 in Figure 1.7b) and divided in half (note however, that while the maximum signal is 300, the baseline is at 20, so the actual signal height is 280); hence, the "half maximum" refers to half the maximum signal. The width of the peak, in time units, is then found at the corresponding $1/2$-height as shown (i.e., at $140+20$, or 160). Naturally, broader peaks have greater values of $W_{1/2}$ and $W_b$ than do narrower peaks, as depicted in Figure 1.8.

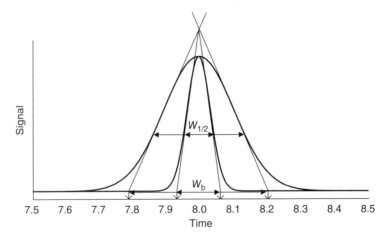

**FIGURE 1.8** Illustration showing that narrower peaks have smaller $W_b$ and $W_{1/2}$ than do broader peaks.

**EXAMPLE** *1.3*

(a) Estimate $W_{1/2}$ and $W_b$ for the peak shown (hint: note that the signal baseline is already at 20).

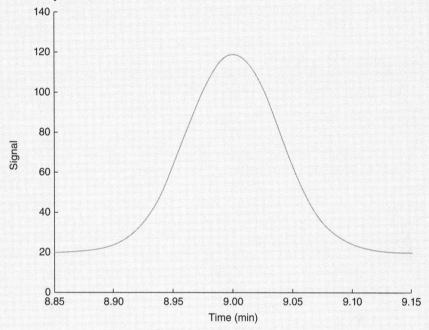

**Answer:**
$W_{1/2} = 0.093$ min, $W_b = 0.16$ min.

The half-width should be measured at a signal of 70 because the total height is 120, but the baseline is at 20. So the actual peak height is 100 signal units tall. Half of this is 50, added to the baseline of 20, indicates that the half-width occurs at a signal strength of 70. The answers given here were obtained by blowing up an image of the peak and estimating the results using a ruler, so values differ slightly depending on the methodology used. The important points are (1) understanding the two measures and how they differ and (2) that the width at half-height is not simply half of the baseline width.

(b) This question is aimed at showing why peak width is important. Imagine a chromatogram with the following four peaks:

| Peak | $t_r$ (min) | $W_{1/2}$ |
|---|---|---|
| 1 | 3.00 | 0.10 |
| 2 | 4.00 | 0.10 |
| 3 | 6.00 | 0.50 |
| 4 | 7.00 | 0.50 |

Notice that peaks 1 and 2 are separated from each other by 1.00 min, as are peaks 3 and 4. Accurately sketch the chromatogram. Which peaks are better separated (i.e., not overlapping), peaks 1 and 2 or peaks 3 and 4? What do you conclude about peak widths and their influence on separation?

**Answer:**
Peaks 1 and 2 are better separated because they have narrower peak widths. While both sets of peaks are separated by the same time difference, wider peaks can overlap, making the peaks poorly resolved and therefore difficult to quantify. This is why we measure peak widths in practice and devote a lot of discussion in the sections below to factors that affect peak widths.

To make comparisons between different columns, the concept of "theoretical plate number," symbolized by $N$, was introduced. The concept of "plates" is borrowed from distillation columns that have actual plates in the interior (see Figure 1.9). A mixture is heated to boiling at the bottom and a temperature gradient develops along the vertical axis of the column, with cooler temperature at the top. The vapor created by the boiling liquid is enriched in the more volatile components. As this vapor mixture rises, it cools and condensese on the plates. Hot vapors rising up through the column reheat the condensed liquid, causing it to vaporize again, further enriching the vapor in the more volatile components. This process of volatilization and condensation occurs continuously. The result is that the more volatile components are increasingly enriched near the top of the column, leaving the higher-boiling species enriched near the bottom. In industrial distillation processes, the vapor exiting the top of the column, now significantly enriched or composed entirely of the most volatile component, can be condensed. As the temperature of the remaining mixture continues to increase, the less volatile components also exit the column and can be collected. Collecting different fractions at different points in time results in separation or at least partial purification of the components. *Columns with more plates, and thus plates that are closer together, produce better separations* that result in purer components being isolated from the original mixture. It should be noted that the fractional distillation columns that are used in chemical laboratories use glass beads, glass protrusions, or other packing material to provide the surface on which the vapors condense as they rise. In this case, the column does not have plates *per se*, but its performance is still described by the concept of "theoretical plates," as discussed in the following in the context of chromatography.

Chromatography columns do not have discrete, individual plates either and do not separate components based on a temperature gradient along the length of the column, but the theory of plates was borrowed as a way to quantify a chromatographic column's ability to separate the chemicals in a mixture and to compare columns to each other.

The number of theoretical plates for a chromatographic column is given by

$$N = \left(\frac{t_r}{\sigma}\right)^2 = 5.54\left(\frac{t_r}{W_{1/2}}\right)^2 \tag{1.11}$$

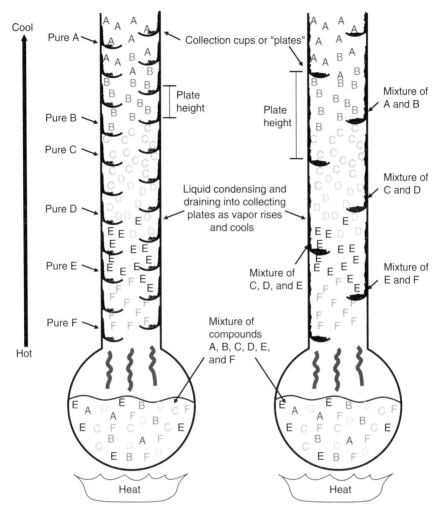

**FIGURE 1.9** A mixture of compounds A, B, C, D, E, and F separated with two different distillation columns. The compound volatility follows the order: volatility of A > volatility of B > ··· > volatility of F. The column on the left has more plates and smaller plate heights (the distance between plates). This leads to a more complete separation of the components of the mixture compared to that achieved with the column on the left with fewer plates and larger plate heights. *Chromatography columns do not have actual, physical plates inside them like the distillation columns pictured here*, but the concept is borrowed as a way to measure and compare the separation ability of different columns. The columns depicted here have around 10–20 actual plates, whereas GC and LC columns have thousands or hundreds of thousands of "theoretical plates."

where $\sigma$ is the standard deviation of the solute peak. The second equality in Equation 1.11 is used because $W_{1/2}$ is easier to measure quickly from a chromatogram than is the standard deviation of the peak. The factor 5.54 assumes that the peak is Gaussian in shape.

Equation 1.11 is essentially a measure of how broad a solute band gets for a given time it spends in the column (hence, $t_r/W_{1/2}$). Columns in which solutes can reside for a long time (large $t_r$) but still produce narrow peaks (small $W_{1/2}$) have high $N$ values. Columns

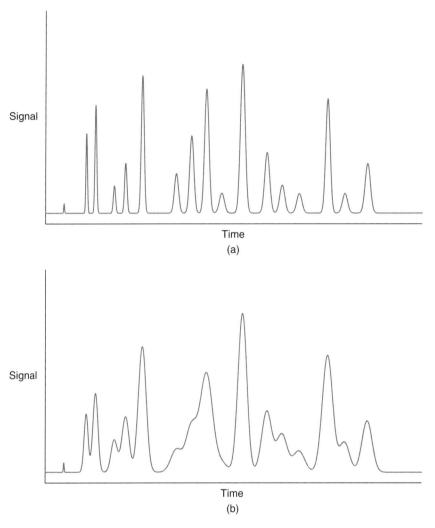

**FIGURE 1.10** Illustration of the effects of broad peaks on peak overlap. The peak maxima occur at the same time in both plots, but the bottom plot has broader peaks resulting from a less efficient column with fewer theoretical plates. It is easier to quantify the peaks in the top chromatogram than in the bottom because in the top, all of the peaks are baseline resolved.

with high plate numbers are favored because more sample components can be completely resolved and quantified (see Figure 1.10).

$N$ is also referred to as the "efficiency" of the column, in addition to being called plate number. If peaks remain narrow even after the solutes have been in the column for a long time, then many components in a mixture can be resolved in a single analysis, providing a highly efficient system.

In practice, the number of theoretical plates is directly proportional to the column length, so long columns naturally have higher $N$ values than shorter columns, all else being equal (i.e., the same stationary phase, column dimensions, particle characteristics). In order to compare columns of different lengths, $L$, a parameter known as the "height equivalent to a theoretical plate (HETP)" or more simply "plate height ($H$)," is introduced:

$$H = \frac{L}{N} \tag{1.12}$$

In terms of plate theory, this relates to the distance, or height, between the collection plates in a distillation column (see Figure 1.9). Smaller plate heights mean more plates for a given column length, resulting in better separations. *Thus, high numbers of theoretical plates (large N) and small plate heights (small H) are associated with narrow peaks* and generally better resolution than columns with smaller $N$ and correspondingly larger plate heights.

Plate theory as developed by Martin and Synge[2] is useful for discussing peak width and providing a way to compare different columns. It also correctly predicts the elution and separation of compounds based on their equilibrium distribution coefficients. At its heart, however, it uses a series of multiple, discrete extraction steps to model chromatography. It also assumes that complete equilibrium is achieved in each step. Chromatography, however, is a continuous process; columns do not have discrete plates, and equilibrium of the solute partitioning between the two phases is never completely established.

Plate theory also ultimately predicts that the solute distribution (i.e., peak width) at any particular point in the column is dictated solely by the column characteristics and is therefore the same for all solutes. This means that the widths of all of the solute zones are predicted to be the same at the end of the column. We know from observation, however, that this is not entirely true. Different solutes elute with different widths on the same column when analyzed under identical conditions. Solute properties, such as their ability to diffuse and their adsorption and desorption kinetics, contribute to peak width. These kinetic effects are not considered in the plate model – a fact explicitly acknowledged by Martin and Synge.[2] So while plate theory is useful in some ways, a different theory known as "rate theory" – which considers these kinetic factors – is used to describe peak broadening. Rate theory, therefore, is used as the basis for our discussion of broadening in the following sections.

## EXAMPLE *1.4*

Suppose the drug warfarin (aka Coumadin) is chromatographed on two different liquid chromatography columns under different conditions and the following data collected:

|          | $t_r$ (min) | $W_{1/2}$ (min) | Column length (cm) |
|----------|-------------|-----------------|--------------------|
| Column 1 | 10.350      | 0.160           | 25.0               |
| Column 2 | 8.721       | 0.130           | 10.0               |

(a) Calculate the number of theoretical plates on both columns (assume Gaussian peak shapes were obtained).

(b) Calculate the HETP (in cm/plate) for both columns.

(c) Which column is better in terms of HETP?

**Answer:**

(a) Plates for column 1:

$$N = 5.54\left(\frac{t_r}{W_{1/2}}\right)^2 = 5.54\left(\frac{10.350}{0.160}\right)^2 = 23{,}182 \approx 23{,}200\,\text{plates}$$

Plates for column 2:

$$N = 5.54\left(\frac{t_r}{W_{1/2}}\right)^2 = 5.54\left(\frac{8.721}{0.130}\right)^2 = 24{,}932 \approx 24{,}900\,\text{plates}$$

(b) HETP for column 1

$$\text{HETP} = \frac{L}{N} = \frac{25.0\,\text{cm}}{23{,}200\,\text{plates}} = 0.00108\,\text{cm/plate}$$

HETP for column 2

$$\text{HETP} = \frac{L}{N} = \frac{10.0\,\text{cm}}{24{,}900\,\text{plates}} = 0.000402\,\text{cm/plate}$$

(c) While the total number of plates is comparable on both columns, because column 2 is shorter, it has a smaller plate height. When comparing columns in terms of HETP values, smaller plate heights are better.

**Another question:**
What is the HETP for conditions under which warfarin elutes at $t_r = 3.715\,\text{min}$, $W_{1/2} = 0.083\,\text{min}$, with a 15-cm column.

**Answer:**
    0.0013 cm/plate.

## 1.2.  BAND BROADENING

The previous section describes ways to quantify peak width and general column performance, but it does not describe the physical processes that lead to band broadening. As mentioned, a model of chromatography known as rate theory is used to examine the movement of molecules in a column and the factors that affect peak width. Rate theory specifically considers four main contributions to band broadening:

1. *Axial* diffusion, also called *longitudinal* diffusion;
2. *Radial* diffusion, also called *lateral* or *transverse* diffusion;
3. The existence of multiple paths with different linear velocities within the column; and
4. The rate of mass transfer (i.e., analyte transport) within and between the stationary and mobile phases, which is influenced by both diffusion and convective (flow) processes.

We develop our understanding of broadening dynamics starting with the simplest system and applying what we learn from it to more complex ones, discussing

1. open tubes with no stationary phase and hence no retention of solutes (not practical for chromatography but a useful place to start);
2. open tubes with a thin stationary phase coating on the walls that causes retention – the situation that exists in capillary gas chromatography; and lastly,
3. packed columns, which applies to virtually all liquid chromatography separations and to gas chromatography separations that are conducted in columns packed with particles coated with a stationary phase (in contrast to capillary columns where the stationary phase is coated on the walls of the columns).

## 1.2.1. Diffusion

An explicit discussion of diffusion is important because it plays a critical role in chromatography. Diffusion is the completely random movement of molecules driven by their translational kinetic energy. This leads to their movement from a region of higher concentration to lower concentration. In other words, solutes diffuse in response to concentration gradients. Diffusion operates in the absence of bulk motion (i.e., no stirring or mechanical mixing).

To picture diffusion and how it relates to band broadening in chromatography, imagine a very narrow plug, or band, of solute molecules that is introduced into an open tube, as shown in Figure 1.11, where the dot density indicates solute concentration. The solute concentration in the areas in front of and behind the band is *initially* zero. Over time, the solute molecules diffuse out from the concentrated plug in both directions due to random molecular motion, as shown in Figure 1.11b and c. The longer the period of time they diffuse, the further they travel. Ultimately, the entire tube will reach the same concentration throughout, assuming it is capped at the ends.

Because diffusion is based on random movements of molecules, Gaussian statistics apply and the broadening of the solute plug over time can be related to the standard deviation in distance units (i.e., the spread of molecules) through Equation 1.13 (the Einstein diffusion equation):

$$\sigma^2 = 2Dt \tag{1.13}$$

where $t$ is time and $D$ is the diffusion coefficient of the diffusing molecules (with units of length$^2$/time). This equation shows that the standard deviation of molecules that diffuse rapidly (higher $D$) is larger than for molecules that diffuse more slowly when given the same amount of time to diffuse. It also shows that broadening increases over time. The effect of time and diffusion coefficients on the spread of molecules is depicted in Figure 1.12.

Diffusion coefficients measure how rapidly the molecules spread out in the medium they are in. Small molecules generally have larger diffusion coefficients than large molecules. Also, diffusion coefficients measured in the gas phase are 10,000–100,000 times greater than in the liquid phase – these are important relationships that should be noted.

It is important to note also that diffusion is rigorously the movement of molecules in response to a concentration gradient. This is in contrast to convection, which is the transport of molecules caused by bulk flow arising from sources such as stirring, pushing liquids and

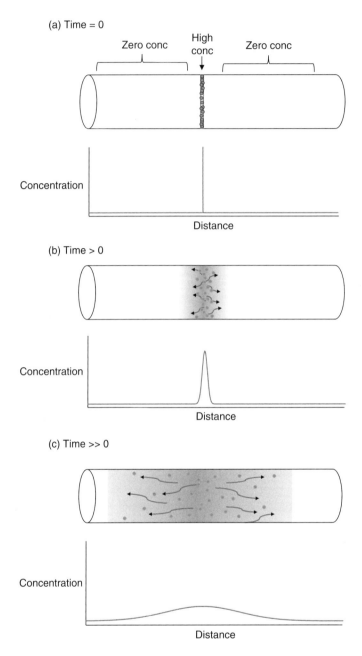

**FIGURE 1.11**  Depiction of solute diffusion over time in an open tube. (a) An infinitely narrow plug of solute molecules. (b, c) Solutes diffuse toward regions of lower concentration down the long axis of the column (i.e., longitudinally). The more time that is allowed for diffusion, the broader the distribution of solute molecules, as depicted below each column.

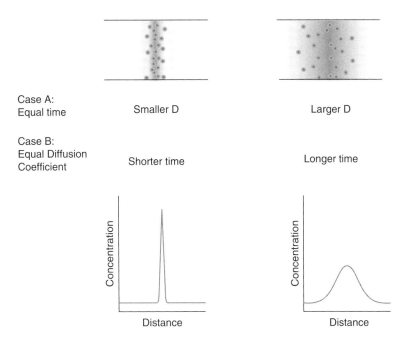

**FIGURE 1.12** The effect of time and diffusion coefficient on band spreading. In case A, the time allowed for diffusion is assumed to be equal. In this case, molecules with higher diffusion coefficients spread out more than do ones with smaller diffusion coefficients. In case B, the diffusion coefficients of the two sets of molecules are assumed to be the same. In this case, the longer the time the molecules are allowed to diffuse, the more broadening occurs. So increased time and higher diffusion coefficients both lead to increased band spreading. Conversely, shorter times allowed for broadening and smaller diffusion coefficients reduce the amount of band broadening.

gases through tubes using higher pressure on one side than the other, or from density gradients. Many of the band-broadening factors we consider below are linked to diffusion, so diffusion coefficients and time are important variables in a number of equations. Diffusion is rather efficient at moving molecules over short distances but bulk flow (convection) is much more effective than diffusion over long distances.

## 1.2.2. Linear Velocity

In this discussion, we talk a lot about velocity, and in all cases we really mean the linear velocity of the mobile phase. It has the same definition here as it does when considering cars, planes, etc. It is simply the distance traveled over the time it took to travel that distance. In chromatography, there are really three distance scales that are important. The first is the length of the column, and the relevant time is the time it takes a completely unretained species to travel the length of the column. The major transport mechanism on this scale is the flow (convection) of mobile phase through the column. So below, when we use the word velocity, we are envisioning the speed with which molecules move when they are in the mobile phase. The second and third distance scales are the radius of the open tube and the diameter of a particle in a packed column. The major transport mechanism on these last two scales is diffusion.

Molecules can travel through the column on different paths. The paths have different velocities, so all of the molecules are not moving at the same speed all the time. These different velocity paths are a source of band broadening because the molecules that move faster than others elute slightly ahead of the rest while those that are on slower paths elute slightly later. We discuss the consequences of these different velocity paths in the following sections.

We also limit our discussion below to situations in which the mobile phase flow is "laminar" as opposed to "turbulent." Laminar flow means that the mobile phase flows with a regular, "nonviolent" motion. For laminar flow, picture a river with a brisk flow but where the water curves around rocks without creating swirls or foam. In contrast, turbulent flow is comparable to a stretch of the river that has rapids.

### 1.2.3. Broadening in Open Tubes with No Stationary Phase and No Retention

As indicated earlier, we first visualize processes occurring in simple open tubes with no stationary phase and no solute retention. In the following section, we add stationary phase considerations, and then we fill the column with particles and examine the contributions of each of these additions to broadening.

***1.2.3.1. Parabolic Flow Profile and Radial Diffusion in the Mobile Phase.*** To begin our considerations, in this section, we examine the effects that parabolic flow and diffusion have on broadening in open tubes. When a solute plug is introduced in an open tube and then pushed down the column using a carrier fluid (gas or liquid), a parabolic flow profile – also called laminar flow – develops inside the column as shown in Figure 1.13. In parabolic flow, molecules at the center of the column move faster than the molecules near the column walls. In fact, molecules right at the walls are in a region of zero velocity (i.e., not moving) and molecules at the center of the column move at twice the average velocity of all the molecules. In three dimensions, the shape of the flow resembles a bullet. As the pressure is increased, the parabolic flow profile becomes more pronounced and the differences in the velocities of the molecules from the center to the walls increase. In the absence of any other molecular motion, the solute zone spreads out across the entire length of the column, resulting in incredibly broad peaks.

But we know from experience that this does not happen, so what keeps the solute zone from being infinitely broad? The answer is radial diffusion. Consider Figure 1.14. Solutes that are at the center of the column and therefore out ahead of the other solutes experience a radial concentration gradient (i.e., a gradient in the direction of the radius of the column). The concentration at the wall is zero, whereas it is high in the center. So molecules diffuse from the center toward the wall. This is called radial diffusion because it occurs in the direction of the radius of the column. In doing so, the molecules move from a faster flow path to a slower path. Solutes at the wall at the rear of the zone also experience a radial concentration gradient. So solutes near the wall diffuse radially toward the center, and in doing so move from a slower flow path to a faster one. This occurs throughout the parabolic flow profile, so molecules are constantly randomly diffusing between different velocity

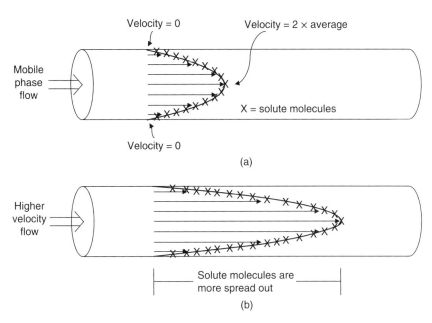

**FIGURE 1.13** Depiction of the parabolic flow profile. (a) Parabolic flow at a lower mobile phase velocity. (b) With a higher mobile phase velocity, the parabolic flow profile becomes more pronounced and solutes are spread out over a greater distance (assuming at this point that there is no mechanism for combating the spread of molecules).

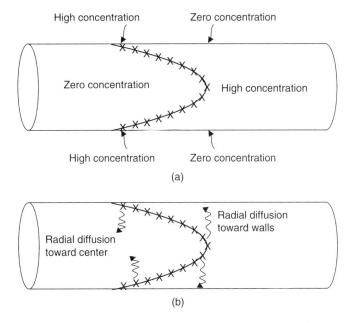

**FIGURE 1.14** Depiction of radial diffusion in response to concentration gradients caused by parabolic flow. X = solute molecules. Note that in (a), on the far right side, the solute concentration is high in the center of the column and zero at the walls. Conversely, on the far left side, the solute concentration is high near the walls and zero in the center of the column. In both cases, a radial concentration gradient exists. In (b), radial diffusion acts to decrease these concentration gradients.

flow paths. Thus, for the entire collection of molecules within the zone, there is an overall *average axial velocity* at which the zone travels through the column.

As with any process dictated by random movements, like diffusion, most of the molecules travel at or close to the average velocity, while a few travel faster and a few slower. The molecules that move slightly faster elute from the column at slightly shorter retention times, those that travel slightly slower elute at longer times, and most of them elute around the center of the peak. This is consistent with the Gaussian shape of chromatographic peaks.

It is important to note that radial diffusion keeps the solute zone together, which decreases broadening, and thus decreases the plate height, $H$. Chromatographically, this is a good thing. In fact, broadening due to parabolic flow and its relaxation by diffusion is given by[3–7]

$$H = \frac{R^2 \bar{u}}{24 D_m} \tag{1.14}$$

where $H$ is the plate height introduced earlier in the chapter (recall that small $H$ values are favorable), $\bar{u}$ is the average linear velocity of the mobile phase through the column (and therefore also of solutes when they are in the mobile phase), $D_m$ is the diffusion coefficient of the solute in the mobile phase (liquid or gas in the case of LC or GC, respectively) and $R$ is the radius of the tube or column.

From this, we see that molecules that diffuse rapidly (larger $D_m$) produce narrower peaks associated with smaller $H$ values than molecules that diffuse more slowly. This is because solutes that diffuse quickly randomly sample more velocity paths, increasing the averaging that takes place and reducing the spread of solute velocities present within the zone. As stated earlier, smaller molecules tend to have higher diffusion coefficients than do larger ones, so something like benzene (single ring) diffuses more rapidly than pyrene (four fused rings) and produces narrower peaks, all else being equal.

From Equation 1.14, we also see that higher mobile phase velocities lead to broader peaks associated with higher plate heights. This arises from two effects:

1. When the zone is moved through the column faster, an exaggerated parabolic flow profile exists, as seen in Figure 1.13, which naturally increases the breadth of the peak.
2. Furthermore, because the solutes are traveling faster, they spend less time in the column. This means there is less time for individual solute molecules to diffuse radially and experience many different velocity paths, so less averaging occurs. If less averaging occurs, a broader range of solute velocities exists and solutes elute over a broader range of time, resulting in broader chromatographic peaks.

Lastly, from Equation 1.14 we see that for open tubes, bigger column radii lead to bigger plate heights (broader peaks), and conversely, smaller radii lead to narrower peaks. Bigger radii mean longer distances that the molecules have to diffuse in order to randomize the velocity paths they experience. With bigger radii, solutes experience fewer velocity paths and thus less randomization and more broadening. For this reason, narrower columns produce smaller $H$ values, and subsequently higher $N$ (theoretical plates) and more efficient separations than do wider bore columns.

### 1.2.3.2. Effects of Longitudinal Diffusion.

While diffusion acting in the radial direction is a good thing in terms of keeping the solute zone together, solute diffusion in the direction of the long axis of the column, also known as longitudinal diffusion, must also be considered. As shown when we first considered diffusion (see Figures 1.11 and 1.12), longitudinal diffusion – also called axial diffusion – spreads out the molecules along the length of the column. This occurs because the solute concentration in front of and behind the solute zone is lower than it is in the center of the zone. In contrast to the effects of diffusion in the radial direction, the longer the molecules remain in the column, and the faster they diffuse, the broader the peak gets due to longitudinal diffusion. In fact, the broadening arising solely from longitudinal diffusion can be computed as[8]

$$H = \frac{2D_{\mathrm{m}}}{\overline{u}} \tag{1.15}$$

where $D_{\mathrm{m}}$ is again the diffusion coefficient of the solute in the mobile phase and $\overline{u}$ is the average mobile phase velocity. It is important to keep in mind that we are considering the situation in which there is no stationary phase and therefore no retention. So the average mobile phase velocity is the only factor dictating the length of time a solute zone spends in the column. Lower velocities result in longer times spent in the column and do so equally for all solutes, regardless of structure. Similarly, higher velocities result in shorter times and also affect all solutes equally. Equation 1.15 shows that at low velocities (long time spent in the column), $H$ is large because the molecules have more time to diffuse along the axis of the column. Also, molecules that diffuse faster (higher $D_{\mathrm{m}}$) produce broader peaks. This makes sense because the faster the molecules diffuse away from the center of the peak, the broader the peak gets.

The two sources of broadening can be added together with the result that

$$H = \frac{2D_{\mathrm{m}}}{\overline{u}} + \frac{R^2\overline{u}}{24D_{\mathrm{m}}} \tag{1.16}$$

for solutes being forced through an *open tube* with *no stationary phase* and *no retention* under conditions in which parabolic flow exists.

## EXAMPLE 1.5

These problems are designed to get at the effects of diffusion as well as the relative contributions of the two terms in Equation 1.16.

What value of $H$ is expected for *n*-octane (as a solute) traveling in hydrogen gas as the mobile phase with an average linear velocity of 60.0 cm/s in a 30.0-m column with a 0.250-mm inner diameter, with no stationary phase? In $H_2$, the diffusion coefficient of *n*-octane is 0.211 cm²/s.[9] These parameters mimic those found in gas chromatography (except that a stationary phase is present when actually doing chromatography). To see the relative contributions of longitudinal (axial) diffusion and the parabolic flow profile, *explicitly calculate the value of the two terms in Equation 1.16 separately.*

**Answer:**

$$H = \frac{2D_m}{\bar{u}} + \frac{R^2\bar{u}}{24D_m}$$

$$H = \frac{2(0.211\,\text{cm}^2/\text{s})}{60.0\,\text{cm/s}} + \frac{\left[\frac{1}{2}\left(0.250\,\text{mm} \times \frac{\text{cm}}{10\,\text{mm}}\right)\right]^2 (60.0\,\text{cm/s})}{24(0.211\,\text{cm}^2/\text{s})}$$

$$H = 0.00703\,\text{cm} + 0.00185\,\text{cm}$$

$$H = 0.00888\,\text{cm}$$

Note that the factor of $\frac{1}{2}$ in the second term is introduced because we have been given the inner diameter of the column but the equation calls for the radius.

It is clear that the longitudinal diffusion term makes a much bigger contribution to broadening than does the parabolic flow profile *in this situation*. But as the average velocity is increased, the first term decreases and the second increases in magnitude.

**Another question:**

(a) To see the effect of diffusion coefficients on $H$, repeat the calculation using the diffusion coefficient of *n*-octane in nitrogen ($0.0460\,\text{cm}^2/\text{s}$).

(b) What effect did the diffusion coefficient have on the contributions to broadening from each of the two terms?

**Answer:**

(a) $H = 0.00153\,\text{cm} + 0.00849\,\text{cm} = 0.0100\,\text{cm}$.

(b) Slower diffusion significantly decreases the broadening arising from the first term (longitudinal diffusion), and significantly increases the broadening due to the second term, which accounts for the broadening due to the parabolic flow profile and its reduction by radial diffusion.

### 1.2.4. Broadening in Open Tubes with a Stationary Phase

When a stationary phase is present, Equation 1.16 for broadening in open tubes gets modified in two ways:

1. The second term on the right-hand side becomes dependent on the amount of retention the solutes experience (i.e., dependent on the retention factor, $k$); and
2. A whole new term gets added to account for slow mass transfer (i.e., diffusion) that occurs in the stationary phase.

The result is the following:

$$H = \frac{2D_m}{\bar{u}} + \frac{1}{24}\left(\frac{1 + 6k + 11k^2}{(1+k)^2}\right)\frac{R^2\bar{u}}{D_m} + \frac{2}{3}\left(\frac{k}{(1+k)^2}\right)\frac{d_f^2\bar{u}}{D_s} \tag{1.17}$$

where $d_f$ is the thickness of the stationary phase film and $D_s$ is the diffusion coefficient of the solute in the stationary phase. This equation, which is referred to as the Golay equation,[10] is often written more simply where $f(k)$ and $f'(k)$ just represent the respective functions of retention factor that they replace:

$$H = \frac{2D_m}{\bar{u}} + \frac{f(k)}{24} \frac{R^2 \bar{u}}{D_m} + \frac{2f'(k)}{3} \frac{d_f^2 \bar{u}}{D_s} = B/\bar{u} + C_m \bar{u} + C_s \bar{u} \tag{1.18}$$

where $B = 2D_m$, $C_m = \frac{f(k)}{24} \frac{R^2}{D_m}$, and $C_s = \frac{2f'(k)}{3} \frac{d_f^2}{D_s}$.

The change in the second term compared to the equation for open tubes without a stationary phase results from the fact that solutes near the wall randomly sorb onto or into the stationary phase and thus temporarily stop moving down the column. In the meantime, solutes that remain in the mobile phase continue down the column as shown in Figure 1.15.

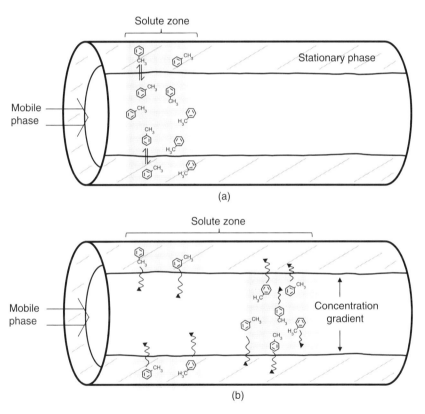

(a)

(b)

**FIGURE 1.15** Effect of radial diffusion in the presence of a stationary phase to help reduce zone broadening. In (a), the solute is in equilibrium between the stationary and mobile phases. In (b), the solutes in the mobile phase have moved down the column, meaning that those in the stationary phase lag behind. This causes the solute zone to broaden. Radial diffusion (signified by the squiggly arrows) in response to the concentration gradients that get created mitigates this effect by averaging out the rate of travel. This process of solutes diffusing into and out of the stationary phase in response to concentration gradients gets repeated the entire time that the solutes spend in the column. Note that the stationary phase thickness in these images has been grossly exaggerated. This has been done simply to help the reader picture the dynamic processes occurring in the column. In reality, the stationary phase thickness is nearly negligible compared to the column diameter.

It is clear that this broadens the solute zone because the molecules are now spread over a greater distance in the column and thus elute over a broader range of time.

The broadening is mitigated by the fact that solutes that move ahead in the mobile phase experience a radial concentration gradient. The gradient is created because some solute molecules that had been in the mobile phase are now interacting with the stationary phase (see Figure 1.16). This means that the solute concentration in the region near the stationary phase is lower than at the center of the column. In response, solutes diffuse radially from the middle toward the stationary phase. By diffusing into this region, these molecules can now also be retained by sorbing to the stationary phase (either *ab*sorbing *into* the stationary phase or *ad*sorbing *onto* the surface). In the meantime, the solutes that had been left behind can move back into the mobile phase and catch up with the zone.

More diffusion radially toward the stationary phase leads to more sorption events per molecule, leading to more averaging of velocities. More averaging leads to narrower peaks and smaller $H$ values. Resistance to mass transfer (i.e., solute motion) therefore leads to broader peaks. For this reason, the second term in Equation 1.17 is often referred to as the "slow mass transfer in the mobile phase" term. As Equation 1.17 shows, if diffusion in the mobile phase is rapid (large $D_m$), solutes diffuse radially faster and increase the opportunity for averaging the retention of all solute molecules. This reduces the broadening that arises from retention caused by the stationary phase.

Thus, the second term in Equation 1.17 incorporates three different effects:

1. broadening due to parabolic flow (i.e., different velocity flow paths);
2. relaxation of broadening associated with the different flow paths by radial solute diffusion; and
3. the need for rapid radial mass transfer in the mobile phase via diffusion to counteract the broadening effects of retention.

The new term in Equation 1.17 compared to Equation 1.16 arises from slow mass transfer in the stationary phase. To understand this, picture a molecule that is absorbed in a thin film of stationary phase with a given thickness, $d_f$. As stated, a retained molecule does not move down the column and therefore temporarily lags behind molecules in the mobile phase. In order to catch up, it must first diffuse through the stationary phase in order to get back to the stationary/mobile phase interface. Once it is at the interface, it can then reenter the mobile phase and resume its journey down the column. The faster it diffuses in the stationary phase, the faster it reaches the interface and the less it lags behind, resulting in a narrower peak. This is consistent with the third term in Equation 1.17, which shows that solutes that diffuse quickly (large $D_s$) produce narrower peaks (smaller $H$). Conversely, slow diffusion in the stationary phase (i.e., resistance to mass transfer) produces broader peaks.

The film thickness also plays a role. It takes a lot less time for molecules to diffuse through a thin film than a thick film. Reducing the time it takes for the solutes to reenter the mobile phase helps keep the solute zone together. Thus, columns with thin films produce narrower peaks than those with thicker films, all else being equal. It should be noted that the film thickness ($d_f$) is squared in the third term, so small changes in $d_f$ can produce large changes in $H$. While thin films are advantageous from a broadening perspective, they also offer less overall retention and less overall solute capacity, so there are legitimate reasons for using thicker films in some circumstances.

The last terms also depend on the mobile phase velocity. Increasing velocity means the solutes spend less time in the column. This means less time for all of the molecules

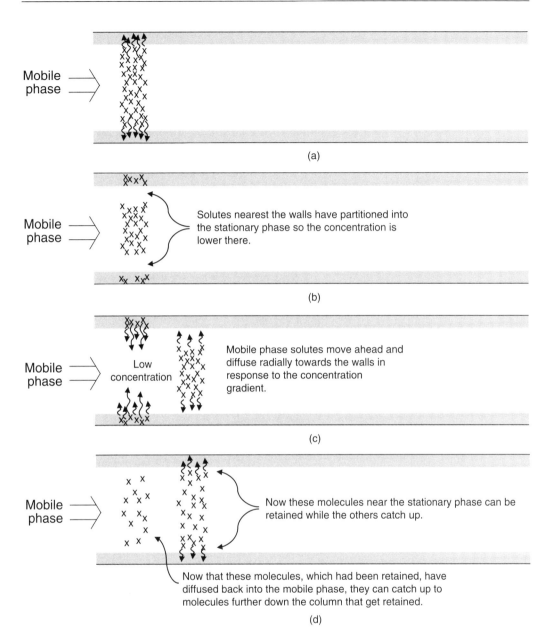

**FIGURE 1.16** An illustration of the process of radial diffusion in response to solute retention caused by the presence of a stationary phase. (a) Solutes in the mobile phase initially partition into the stationary phase. (b) Solutes nearest to the stationary phase are those most likely to partition. When they do, a radial solute concentration exists in the mobile phase, with high solute concentration in the center of the column and low concentration near the walls. (c) Solutes in the mobile phase continue down the column and diffuse to reduce the radial concentration gradient. Similarly, solutes in the stationary phase diffuse out in response to the low concentration in the center of the column that arises because the mobile phase has pushed the rest of the molecules down the column. (d) The molecules that have reentered the mobile phase will now be pushed down the column, while molecules out in front are retained as they approach and interact with the stationary phase. It is important to note that these are static pictures, but in reality, all of these processes, in addition to parabolic flow and longitudinal diffusion, are happening in the column continuously and simultaneously. But breaking them up into discrete steps makes it easier to understand the effects of each process on band broadening.

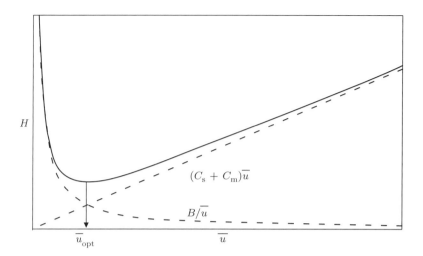

**FIGURE 1.17** $H$ versus $\bar{u}$ plot for an open tube with a thin stationary phase. There are three main contributions: the $B$, $C_s$, and $C_m$ terms as shown in Equation 1.18. The $C_s$ and $C_m$ terms are both linear with $\bar{u}$ and their contribution to $H$ can be combined as shown. The contribution to $H$ of the $B/\bar{u}$ term approaches zero at high average linear velocities. The solid black line is the sum of the two contributions, or the total $H$. Note that at high average linear velocities, the $(C_s + C_m)$ term makes the vast majority of the contribution to the overall $H$. Conversely, at low average linear velocities, the $B/\bar{u}$ term dominates. The arrow indicates the optimum average linear velocity ($\bar{u}_{opt}$), meaning the velocity that produces the lowest plate height and therefore the narrowest peaks. As noted in the text, however, we frequently operate at linear velocities higher than the optimum, accepting the slightly broader peaks that result in order to decrease the total analysis time.

to experience multiple adsorption events and thus less averaging of velocities. So as $\bar{u}$ increases, broadening – as measured by $H$ – does too.

It is worth stopping here to point out that the first term in Equation 1.17 indicates that high velocities (associated with high mobile phase flow rates) are favored to reduce $H$, whereas the second and third terms indicate that low velocities are favorable. Combined, the three terms yield an $H$ versus $\bar{u}$ plot (Figure 1.17) that indicates that for any column or tube there is an optimal velocity ($\bar{u}_{opt}$) that produces the narrowest peaks (smallest $H$). While it might seem logical to operate at this optimal velocity, in practice we generally operate at higher velocities and accept a slight increase in broadening for the sake of increased speed and faster analyses. This is discussed in more detail later in the chapter.

## EXAMPLE *1.6*

These problems build on the calculation in Example 1.5 to explore the effect of retention and slow mass transfer in the mobile phase on broadening that were just discussed.

We will use the same conditions as in **Example** 1.5:

- *n*-Octane as the solute with $D_m = 0.211$ cm$^2$/s
- Hydrogen gas as the mobile phase

- Average linear velocity of 60.0 cm/s
- Column: 0.250-mm inner diameter, 30-m long.

Now, in this problem, let there be a stationary phase coated on the wall that is 0.25-μm thick and let the retention factor for *n*-octane be 3.25. Let the diffusion coefficient of *n*-octane in the stationary phase be $1.20 \times 10^{-5}$ cm$^2$/s (note how much lower this is compared to the value in the gaseous hydrogen mobile phase). Calculate the HETP that results from these conditions. Note that these parameters reasonably mimic those found in gas chromatography.

*Suggestion*: Explicitly calculate the value of the three terms in Equation 1.17 separately and compare them to the related terms in Example 1.5. This will be more informative and interesting and help provide a sense of the relative importance of the three different terms.

**Answer:**

$$H = \frac{2D_m}{\overline{u}} + \frac{1}{24}\left(\frac{1 + 6k + 11k^2}{(1+k)^2}\right)\frac{R^2\overline{u}}{D_m} + \frac{2}{3}\left(\frac{k}{(1+k)^2}\right)\frac{d_f^2\overline{u}}{D_s}$$

$$H = \frac{2(0.211\,\text{cm}^2/\text{s})}{60.0\,\text{cm/s}} + \frac{1}{24}\left(\frac{1 + 6(3.25) + 11(3.25)^2}{(1 + 3.25)^2}\right)\left(\frac{(0.0125\,\text{cm})^2 60.0\,\text{cm/s}}{0.211\,\text{cm}^2/\text{s}}\right)$$

$$+ \frac{2}{3}\left(\frac{3.25}{(1 + 3.25)^2}\right)\left(\frac{\left(0.250\,\mu\text{m} \times \frac{10^{-4}\,\text{cm}}{\mu\text{m}}\right)^2 60.0\,\text{cm/s}}{1.20 \times 10^{-5}\,\text{cm}^2/\text{s}}\right)$$

$$H = 0.00703\,\text{cm} + (7.57)0.00185\,\text{cm} + 0.000375\,\text{cm}$$

$$H = 0.00703\,\text{cm} + 0.0140\,\text{cm} + 0.000375\,\text{cm}$$

$$H = 0.0214\,\text{cm}$$

The factor of 7.57 in the second term is shown explicitly to show the impact that solute retention has on broadening. Solutes in the stationary phase are not moving down the column, while the solutes that are in the mobile phase move ahead of them. The higher the retention factor, the bigger the effect.

Comparing these results to Example 1.5 shows that the first term due to longitudinal diffusion is the same. The second term, which incorporates the effects of retention and slow mass transfer in the mobile phase, is now the largest contributor to broadening, and slow mass transfer in the stationary phase adds a whole new term but not a significant one because the given film thickness is quite thin.

**Another question:**

(a) Now, as you did in Example 1.5, repeat the calculation but with nitrogen as the mobile phase. Recall that the diffusion coefficient of *n*-octane in $N_2$ was given as $0.0460$ cm$^2$/s.

(b) By comparing the results in this example, which represent actual GC practice fairly well, which mobile phase is better, $H_2$ or $N_2$? Why?

**Answers:**

$$H = \frac{2D_m}{\overline{u}} + \frac{1}{24}\left(\frac{1+6k+11k^2}{(1+k)^2}\right)\frac{R^2\overline{u}}{D_m} + \frac{2}{3}\left(\frac{k}{(1+k)^2}\right)\frac{d_f^2\overline{u}}{D_s}$$

$$H = \frac{2(0.0460\,cm^2/s)}{60.0\,cm/s} + \frac{1}{24}\left(\frac{1+6(3.25)+11(3.25)^2}{(1+3.25)^2}\right)\left(\frac{(0.0125\,cm)^2 60.0\,cm/s}{0.0460\,cm^2/s}\right)$$

$$+ \frac{2}{3}\left(\frac{3.25}{(1+3.25)^2}\right)\left[\frac{\left(0.250\,\mu m \times \frac{10^{-4}\,cm}{\mu m}\right)^2 60.0\,cm/s}{1.20\times10^{-5}\,cm^2/s}\right]$$

$$H = 0.00153\,cm + (7.57)0.00849\,cm + 0.000375\,cm$$

$$H = 0.00153\,cm + 0.0643\,cm + 0.000375\,cm$$

$$H = 0.0662\,cm$$

(a) $H = 0.0662$ cm.
(b) $H_2$ is better because it leads to smaller plate heights. This is due to the higher diffusion coefficient of the solute in $H_2$, which means solute molecules diffuse radially toward the stationary phase faster than they do in nitrogen. This reduces the broadening caused by retention in $H_2$ more than in $N_2$ as demonstrated by the relative magnitudes of the second terms. This counteracts and outweighs the decrease in the longitudinal diffusion contribution (i.e., the first term).

***1.2.4.1.   A Note About Diffusion Before Considering Packed Columns.*** In the treatment above, we account for broadening of the solute zone caused by solute diffusion in the mobile phase along the axis of the column, but not by solutes retained in the stationary phase. Molecules do, in fact, also diffuse longitudinally while in the stationary phase. However, we typically apply Equation 1.17 to capillary gas chromatography. In this particular case, the broadening caused by longitudinal diffusion in the stationary phase is negligible compared to that in the mobile phase because diffusion coefficients in the gas phase are 10,000–100,000 times greater than in condensed, or liquid-like phases (such as the polymeric phases used in GC). Thus, broadening due to longitudinal diffusion in the stationary phase is negligible compared to that in the mobile phase in GC. This is not so in LC and modifications to Equation 1.17 are required to account for longitudinal diffusion effects in the stationary phase.[11]

### 1.2.5.   Broadening in a Packed Column

Capillary gas chromatography is generally conducted in open, narrow tubes and therefore follows the broadening phenomena discussed above. But some gas chromatography, and all of liquid chromatography, is conducted in columns packed with particles coated with a stationary phase. While many of the sources of broadening discussed above are still at work in packed columns, the presence of the packing requires modifications to the $H$ versus $\overline{u}$

equation. Using what is known as the "random walk" model of chromatography, Giddings developed the following equation[12]:

$$H = \frac{2\gamma D_{\mathrm{m}}}{\bar{u}} + \sum \left( \frac{1}{2\lambda_i d_{\mathrm{p}}} + \frac{D_{\mathrm{m}}}{\omega_i d_{\mathrm{p}}^2 \bar{u}} \right)^{-1} + q \frac{k}{(1+k)^2} \frac{d^2 \bar{u}}{D_{\mathrm{s}}} \qquad (1.19)$$

The first and third terms on the right-hand side are essentially identical to those in the Golay equation (Equation 1.17) and arise from the same fundamental phenomena. The middle term therefore receives the bulk of our attention. So in the following sections, we briefly address the first and third terms, followed by a more detailed discussion of the second term.

**1.2.5.1.  The First Term.** The first term $\left( \frac{2\gamma D_{\mathrm{m}}}{\bar{u}} \right)$ accounts for the broadening arising from longitudinal diffusion in the mobile phase. Compared to this term in the Golay equation, an additional factor, $\gamma$, is included. This is referred to as the obstruction factor and accounts for the fact that solid particles block or disrupt the diffusion of molecules down the axis of the column. Because of this, solutes do not diffuse as far away from the center of the zone as they would in an open tube. In a typical column, $\gamma$ is between 0.6 and 0.8, whereas $\gamma = 1.0$ for an open tube. Inserting a value of less than 1.0 into the first term decreases its magnitude, resulting in a decrease in $H$, corresponding to narrower peaks in packed columns compared to open tubes when considering this term only.

More importantly, $D_{\mathrm{m}}$ is the diffusion coefficient of the solute in the mobile phase, which, in the case of LC, is a liquid. As noted earlier, diffusion coefficients in liquids are 10,000–100,000 times lower in liquids than in gases.

**1.2.5.2.  The Third Term.** The third term $\left( q \frac{k}{(1+k)^2} \frac{d^2 \bar{u}}{D_{\mathrm{s}}} \right)$ in the packed column equation (Equation 1.19) is identical to the third term in the open tube equation except for the factor $q$. This factor accounts for the fact that in a packed column, the stationary phase is bonded to or coated on the surface of the pores of the particle and thus is not a uniformly thin film. In the case of regular films coated on a wall, $q = 2/3$ as seen in the Golay equation. With less regular geometry, $q$ needs to be determined experimentally, but its actual value is of little importance here, where our focus is on visualizing the sources of band broadening and the dynamics that occur inside a chromatography column.

Another change is that $d_{\mathrm{f}}$ has been replaced by simply $d$, which is a measure of the depth of the stationary phase through which the solute can diffuse. If the stationary phase is coated on hard particles, $d$ is essentially the film thickness. In developing this equation, however, Giddings tried to be as general as possible, envisioning not only solid spheres but also allowing for particles that have pores that are completely filled with stationary phase. With such particles, $d$ represents the depth of the pores and the value of $q$ depends on their shape (conical, cylindrical, etc.). In all cases, however, regardless of the specific mathematical form, the third term accounts for the broadening due to solute retention and resistance to mass transfer in the stationary phase.

**1.2.5.3.  The Second Term.** To understand the middle term in Equation 1.19, it is necessary to consider the velocity profile in a packed column. We saw that in open tubes, the velocity varies regularly from the wall to the center of the column. The only mechanism

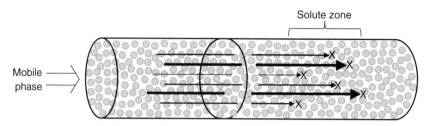

**FIGURE 1.18** Velocities of the mobile phase vary in different regions of a column packed with particles (gray dots). The different thicknesses of the arrows represent the differing velocities. The existence of multiple flow paths, along which solutes are carried at different velocities, is a source of band broadening in packed columns. This is illustrated by the solutes (X) at the front of each arrow. They are spread out in their location within the column because of the velocity variations. This means that they will also elute at slightly different times, resulting in band broadening. Note: The size of the particles is greatly exaggerated relative to the diameter of the column. Recall that the most common particle sizes are 5 µm and smaller.

by which an individual molecule can change its velocity is by diffusing radially into a different velocity path. The situation in packed columns is more complicated because the velocity profile across the column is irregular due to the presence of the particles. Velocities are higher along some paths and lower along others, depending on how the particles are packed together – more rigorously, the "packing density" – in different regions of the column. This is depicted in Figure 1.18.

The existence of different velocity paths creates broadening because molecules that reside in faster paths elute before those in slower paths, as shown in Figure 1.18. This phenomenon is quite comparable to the different velocities present in parabolic flow in open tubes, except that the velocities do not vary in a regular way across a packed column as they do in an open tube. The broadening created by the velocity irregularities due to the presence of the particles is often referred to as eddy dispersion or eddy diffusion.

### 1.2.5.4. Mechanisms That Reduce Broadening.
Solute diffusion between different velocity paths is the only mechanism that counteracts the broadening due to radial velocity variations in open tubes. Two mechanisms work in tandem in packed columns to reduce the effects of different velocity paths:

1. Changes in velocity within a given path; and
2. Diffusion of solute molecules between different paths.

Consider Figure 1.19 that shows two different paths that molecules can take through a region of a packed column. A molecule on Path 1 initially moves slowly because the particles are packed tightly together. The velocity in that region will be low due to the obstruction caused by densely packed particles, but the packing "opens up" down the column and the velocity increases in the more open region. On the other hand, molecules on Path 2 initially move at a higher velocity but then slow down where the particles are packed more closely together.

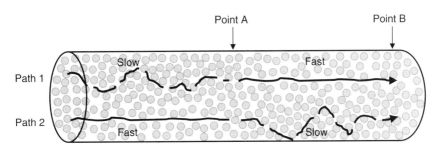

**FIGURE 1.19**  Depiction of the "flow mechanism" of broadening relaxation. A solute on Path 1 initially experiences a slow velocity due to a high packing density in a particular region of the column. As the packing density decreases, it then experiences a faster velocity. Conversely, a solute on Path 2 initially experiences faster velocity, but then slows as it enters a region of higher packing density. The solute on Path 2 reaches Point A first, but the two solutes might still elute close together in time due to the averaging of velocities that they experience as the move through the column. Again, the size of the particles is exaggerated relative to the column dimensions.

It is clear that solute molecules on Path 2 will get to Point A in the column ahead of solute molecules on Path 1. However, it is equally clear that as the mobile phase and the solute molecules in it on Path 2 slow down between Points A and B, molecules on Path 1 speed up. So the *average* velocity of solute molecules on the two paths might be quite comparable. As a result, solutes on the two paths might elute quite close in time to each other despite the existence of different velocities within the column. Furthermore, note that in this situation, the averaging of the velocities does not occur because of diffusion. In fact, in this extreme example, the molecules never leave their flow paths. The averaging results from variations in the velocity at different axial positions within the flow paths themselves.

There are clearly more than two flow paths through a column, and in fact some paths intersect with other paths, but the general principle of random variation of the velocities of the paths down the length of the column provides a mechanism for averaging out the velocities experienced by the solute molecules. This reduces, but does not entirely eliminate, the broadening that occurs because of the presence of multiple flow paths. Giddings called this mechanism of broadening relaxation the *"flow mechanism."*[13]

The second mechanism that causes averaging of velocities is one we have already seen – radial diffusion of solutes through the mobile phase from one path to another. This

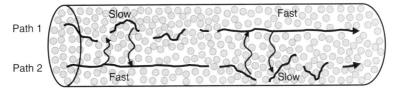

**FIGURE 1.20**  Depiction of the "diffusion mechanism" of broadening relaxation. Solutes diffuse in the radial direction (depicted by arrows). In doing so, solutes that had been in fast paths move into slower ones, and conversely, solutes that had been in slow paths move into faster ones. In this way, the velocities experienced by all solutes average out, decreasing the effects of band broadening caused by the existence of different velocity paths through a packed column.

is depicted in Figure 1.20 and is called the *"diffusion mechanism."* Just as in parabolic flow, molecules diffuse radially from one flow path to another and in this way experience many different velocities. The more the molecules do this, the more they tend to travel at or near the average velocity and thus elute in a narrow band.

In addition, we saw in the open tube case that diffusion through the mobile phase also brings solutes to the stationary phase where they can be retained. Slow solute diffusion (i.e., slow mass transfer) in the mobile phase toward the stationary phase reduces the number of times a solute is retained. This limits the amount of averaging of the velocities that occurs over all the solutes, leading to a wider distribution of velocities and broader peaks.

### 1.2.5.5.  *Mathematical Description of the Flow and Diffusion Mechanisms.* In this section, we will first look at two extremes:

1. The case in which the flow mechanism is the only mechanism combating the broadening due to multiple flow paths; and
2. The case in which the diffusion mechanism is the only mechanism at work.

The first case dominates at high average velocities. At high velocities, the molecules spend little time in the column and thus have very little time to diffuse between flow paths, essentially eliminating the potential for the diffusion mechanism to reduce broadening. In contrast, the diffusion mechanism dominates at low velocities, where the molecules have extensive time to experience all of the various velocities. In this case, the reduction in broadening due to diffusion swamps any reduction arising from the flow mechanism.

After having established these two extremes, we address the more realistic situation in which both mechanisms are acting in parallel.

### 1.2.5.6.  *Case 1 – Flow Mechanism Only.* If the flow mechanism is the dominant cause of the solute velocity averaging, then the contribution to plate height arising from different velocity paths is given by

$$H_f = 2\lambda d_p \tag{1.20}$$

where the subscript "f" indicates the flow mechanism, $d_p$ is the diameter of the particles used to pack the column, and $\lambda$ is a constant related to the range of velocities present relative to the average velocity ($\Delta u/\overline{u}$) and to the packing structure within the column. The $\lambda$ parameter characterizes how well packed a column is, with smaller values being associated with narrower peaks and smaller plate heights. It is also important to note that $H$ is dependent on $d_p$, with smaller particles leading to narrower peaks (lower $H$ values). This is one of the reasons manufacturers of chromatography columns and instruments have continuously tried to produce systems that can use smaller and smaller particles.

### 1.2.5.7.  *Case 2 – Diffusion Mechanism Only.* If the diffusion mechanism is the only mechanism reducing the broadening caused by the presence of different velocity paths, then the contribution to $H$ from the different paths is given by

$$H_d = \frac{\omega d_p^2 \overline{u}}{D_m} \tag{1.21}$$

where the subscript "d" indicates the diffusion mechanism, and $\omega$ depends on the range of velocities present, the average velocity, and the packing structure (similar but not identical to $\lambda$ in the equation above).

Here again, we see that smaller particles lead to smaller $H$ values associated with narrower peaks. In fact, the particle size is squared, so using smaller particles leads to a significant reduction in broadening.

This term is also dependent on the mobile phase velocity for the same reasons discussed above in association with broadening in open tubes. As the velocity increases, less time is allowed for the solutes to diffuse between and experience multiple velocities. Therefore, less averaging occurs and solutes elute over a larger range of times.

Equation 1.21 also shows that solutes that diffuse faster (higher $D_m$) have narrower peaks than those that diffuse more slowly. This makes sense because rapidly diffusing molecules transfer between different velocity paths more frequently, thus experiencing more velocities, leading to more averaging of solute velocities throughout the column and concomitantly narrower peaks.

### 1.2.5.8. Combining the Flow and Diffusion Mechanisms – The Coupling Term.

While in the preceding sections we looked at two separate mechanisms for reducing broadening due to multiple flow paths, it is typical for both mechanisms to be operating at the same time – that is, in parallel. It might be intuitively appealing to think that when they are both contributing we simply add the two equations together. This idea of linear additivity, however, does not apply because the two mechanisms are not independent of each other. For example, a radial diffusion event can cut short, or eliminate, a molecule's experience of a random change in velocity within a given velocity path, and similarly, a random change in velocity along a path can preempt a solute's diffusion into a different velocity path.

When both mechanisms are acting, the combined contribution to the plate height from the existence of different velocity paths is given by

$$H_c = \frac{1}{\dfrac{1}{2\lambda d_p} + \dfrac{D_m}{\omega d_p^2 \overline{u}}} = \left( \frac{1}{2\lambda d_p} + \frac{D_m}{\omega d_p^2 \overline{u}} \right)^{-1} \tag{1.22}$$

where the subscript "c" stands for "combined" or "coupled." (As an aside, we note that this equation mirrors that for determining the net resistance of two resistors in parallel.)

### 1.2.5.9. Complete Equation for H in a Packed Column.

Combining all of the mechanisms that lead to broadening in packed columns leads to Equation 1.23:

$$H = \frac{2\gamma D_m}{\overline{u}} + \left( \frac{1}{2\lambda d_p} + \frac{D_m}{\omega d_p^2 \overline{u}} \right)^{-1} + q \frac{k}{(1+k)^2} \frac{d^2 \overline{u}}{D_s} \tag{1.23}$$

which is nearly but not quite the same as Equation 1.19 (the second term is the sum over "$i$" processes in Equation 1.19), which was introduced at the start of the packed column discussion.

For the sake of completeness, we note that our discussion above focused on broadening arising from different velocity paths caused by differences in obstruction and openness of the paths. Giddings describes four additional sources of velocity differences within packed columns, each contributing to zone broadening in a manner that is mathematically similar to that discussed above for the middle term.[12] These effects sum together, so Equation 1.23 can be recast more broadly as

$$H = \frac{2\gamma D_m}{\bar{u}} + \sum \left( \frac{1}{2\lambda_i d_p} + \frac{D_m}{\omega_i d_p^2 \bar{u}} \right)^{-1} + q \frac{k}{(1+k)^2} \frac{d^2 \bar{u}}{D_s} \tag{1.24}$$

which is the equation presented at the start of the packed column discussion, in which the summation is taken over all five broadening mechanisms (i.e., $i = 1$–5) identified by Giddings.

It is common to see this equation written more simply as

$$H = \frac{B}{\bar{u}} + \sum \left( \frac{1}{A_i} + \frac{1}{C_{m,i} \bar{u}} \right)^{-1} + C_s \bar{u} \tag{1.25}$$

where

$$A_i = 2\lambda_i d_p \tag{1.26}$$

$$B = 2\gamma D_m \tag{1.27}$$

$$C_{m,i} = \frac{\omega_i d_p^2}{D_m} \tag{1.28}$$

$$C_s = q \frac{k}{(1+k)^2} \frac{d^2}{D_s} \tag{1.29}$$

This equation is often presented in a simplified, approximate form and is known as the van Deemter equation.[14]

$$H = A + \frac{B}{\bar{u}} + C_m \bar{u} + C_s \bar{u} \tag{1.30}$$

or

$$H = A + \frac{B}{\bar{u}} + C \bar{u} \tag{1.31}$$

Table 1.2 summarizes the physical effects that are accounted for by each term.

In open tubes, $A = 0$ because there are no particles, and in situations in which the diffusion mechanism dominates the second term of Equation 1.25 and Equation 1.30 becomes

$$H = \frac{B}{\bar{u}} + C_m \bar{u} + C_s \bar{u} \tag{1.32}$$

which we saw earlier in Equation 1.18. This equation particularly applies to gas chromatography. In capillary GC with open tubes, there are no particles, and in GC with packed

**TABLE 1.2   Physical Significance of the Terms in the van Deemter Equation**

| Term | Accounts for broadening due to |
|------|-------------------------------|
| $A$ | Multiple velocity paths in packed columns (also called eddy diffusion or eddy dispersion) |
| $B$ | Longitudinal diffusion (also called axial diffusion) |
| $C_m$ | Slow radial mass transfer of solutes in the mobile phase between different velocity paths and toward the stationary phase. Also accounts for broadening from the parabolic flow profile in open tubes |
| $C_s$ | Slow mass transfer in the stationary phase allowing solutes in the mobile phase to advance down the column ahead of retained solutes |

columns, the condition that the diffusion mechanism dominates normally applies because the diffusion coefficients in gases are so high.

When the flow mechanism dominates (high velocity situations), the equation can be written as

$$H = A + \frac{B}{\bar{u}} + C_s \bar{u} \tag{1.33}$$

which is consistent with the form of the van Deemter equation (Equation 1.30).

### *1.2.5.10.   Specific Considerations for Modern Liquid Chromatography.*

As mentioned earlier, when Giddings developed his treatment of band broadening that we presented in the previous sections, he largely had in mind GC in wall-coated open tubes or columns packed with particles coated with a stationary phase. Many of the underlying phenomena such as diffusion in the mobile and stationary phases are also relevant to LC, but Equation 1.25 as formulated is only a rough approximation to the band-broadening behavior in liquid chromatography as it is practiced today. The most significant difference is that LC is commonly performed using particles that are porous rather than being completely solid (see Figure 1.21).

The pores in the particle are quite small (60–120 Å pores are typical) and mobile phase does not flow through them even though an elaborate network of channels connect the pores throughout the particle. The pores are, however, filled with the mobile phase that is stagnant inside the pores. This nonmoving mobile phase is therefore referred to as the stagnant mobile phase to differentiate it from the mobile phase that flows around the particles. Both phases, however, have the same composition. The pore walls are also coated with a stationary phase. In the case of RPLC, the stationary phase is typically composed of long alkyl chains chemically bonded to the surface of the pores (see Figure 1.21).

As depicted in Figure 1.21, for solutes to be retained in LC with porous particles,

1. solute molecules must first be brought to the solid support particle through convection (i.e., flow) and diffusion;
2. then they must diffuse into the stagnant mobile phase inside a pore (so solute diffusion coefficients in the mobile phase, $D_m$, are important);
3. then they must diffuse through the stagnant mobile phase (also dependent on $D_m$) to the stationary phase where they can be retained by the stationary phase;

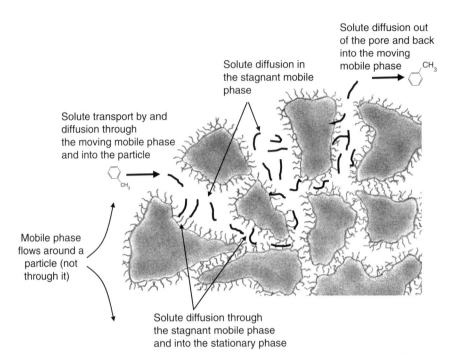

Solute diffusion out of the pore and back into the moving mobile phase

Solute diffusion in the stagnant mobile phase

Solute transport by and diffusion through the moving mobile phase and into the particle

Mobile phase flows around a particle (not through it)

Solute diffusion through the stagnant mobile phase and into the stationary phase

**FIGURE 1.21** A depiction of the events involved in solute retention. A quadrant of a cross section of a particle is shown. Shaded areas represent the silica support particle. White spaces represent the pores, which are filled with stagnant mobile phase. In order to be retained, solutes in the flowing mobile phase must (i) be brought to the solid support particle, (ii) diffuse into the particle pores by diffusing through the stagnant mobile phase that fills the pores, and (iii) diffuse into the stationary phase. The solute must diffuse back through the stagnant mobile phase and out of the pores in order to get back into the flowing mobile phase. When they do, they catch up to other molecules of the same kind that have moved down the column with the mobile phase.

4. there they interact with and diffuse through the stationary phase for some time (this is dependent on $k$ and $D_s$); and
5. finally, the molecules have to diffuse out of the stationary phase back into the stagnant mobile phase, diffuse through the pores in the stagnant mobile phase to a pore exit, and diffuse out of the particle back into the moving mobile phase, all in order to catch up with those molecules that had remained in the flowing mobile phase while they were in the particle.

It is reasonable to deduce that the faster the solutes diffuse in the mobile and stationary phases, the less zone broadening occurs due to the time solutes spent inside the particles. Not surprisingly then, specific mathematical formulations that account for these phenomena include terms related to $D_s$ and $D_m$ and have been treated in the literature.[11,15,16] However, a discussion of them is beyond the scope of this chapter. We do note, however, that Knox and Scott suggested that experimental data acquired over a very wide range of velocities fit equations better if the $A$-term is allowed to be weakly dependent on velocity as well.[11] They therefore modified the van Deemter equation as

$$H = A\bar{u}^{-0.33} + \frac{B}{\bar{u}} + C\bar{u}$$ (1.34)

This equation provides quite good fits to LC data in particular.

### 1.2.6. Putting It All Together

The overall point of the discussion above is that a lot is going on as solute molecules travel through a chromatographic column. The molecules are carried down the column by the flow of the mobile phase. The mobile phase velocity they experience varies from point to point depending on the path they take, and the existence of different velocities is a major source of broadening. Radial diffusion of solutes through the mobile phase reduces the broadening to some extent. These effects are incorporated in the $A$-term for packed columns and in the $C_m$-term for packed and open tubes.

Diffusion of solutes through the mobile phase also transports them to the stationary phase where they can be retained. In capillary GC, this diffusion is toward the coated wall of the column, and in LC with columns packed with porous particles, this diffusion is through both the moving mobile phase and through the stagnant mobile phase inside the pores, toward the stationary phase bonded to the particles. These diffusion processes allow molecules to experience multiple retention events and thus reduce the severe broadening that retention would otherwise cause. These effects are incorporated in the $C_m$-term in the equations presented above.

In addition, diffusion of the retained molecules through the stationary phase helps the retained molecules get back to the stationary/mobile phase interface. Once at the interface, they can move back into the mobile phase and begin to catch up to the solutes that have moved down the column in the mobile phase, which themselves may have been retained as others are catching up. The effects of solute diffusion through the stationary phase are incorporated in the $C_s$-terms in the equations presented above (e.g., Equations 1.17–1.19, 1.23–1.25, and 1.30–1.33).

For the effects just described, the slower the mobile phase velocity, the better in terms of broadening. Slower velocities allow more time for radial diffusion, and more sampling or averaging of the various velocity paths that solutes experience. For these effects, then, lower velocities lead to narrower peaks (desirable) and higher velocities lead to broader peaks (undesirable).

However, we also saw that longitudinal diffusion down the long axis of the column also broadens peaks. The slower the mobile phase velocity, the worse the broadening gets. Thus, when considering only this effect, higher velocities are favored over lower velocities. The effect of longitudinal diffusion is incorporated in the $B$-term in the equations above.

Table 1.3 examines the effects of decreasing the key variables embedded in the $A$, $B$, and $C$ terms and the velocity regime (high versus low) in which they are more important. In all cases, increasing instead of decreasing the key variables reverses the effects described in the table.

**TABLE 1.3   Factors That Impact Band Broadening**

| Factor | Impact | Term impacted | Especially important at | Reasons |
|---|---|---|---|---|
| Smaller $D_m$ | Smaller $H$ | $B$-term | Low mobile phase velocities | Decreased broadening due to longitudinal diffusion |
|  | Larger $H$ | $C_m$-term | High mobile phase velocities | Increased broadening due to less averaging of velocities because of fewer radial transfers between different mobile phase velocity paths. Also less averaging of retention events due to slow transport of solutes to the stationary phase |
| Smaller $D_s$ | Larger $H$ | $C_s$-term | High mobile phase velocities | Increased broadening due to slower transfer of solutes through the stationary phase and back to the mobile phase allows molecules in the mobile phase to move further ahead down the column. High velocities exacerbate the distance between molecules in the stationary and mobile phases |
| Smaller $d$ or $d_f$ | Smaller $H$ | $C_s$-term | High mobile phase velocities | Decreased broadening because solutes in thin stationary phases can diffuse through the thin layer and reenter the mobile phase faster than solutes in thicker stationary phases |
| Smaller $d_p$ | Smaller $H$ | $A$-term | High mobile phase velocities | The presence of particles creates multiple velocity paths in the column. The velocity within a path varies with longitudinal distance along the column, and the length of the variations (slow versus fast segments within a particular stream) scales with particle size. This "flow mechanism" term becomes more important at high mobile phase velocities relative to the "diffusion mechanism" term because molecules do not have enough time to diffuse radially between flow paths to average out the velocities |
|  | Smaller $H$ | $C_m$-term | High mobile phase velocities | Decreased broadening because smaller particles lead to shorter radial distances that solutes have to diffuse in order to join a different velocity path, so more averaging of velocities occurs |

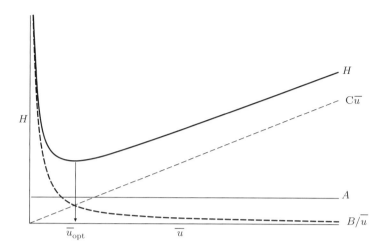

**FIGURE 1.22** $H$ versus $\bar{u}$ plot for packed columns. There are three main contributions: the $A$-, $B$-, and $C$- terms as shown in Equation 1.33. The $A$-term accounts for multiple velocity paths, the $B$-term accounts for longitudinal diffusion, and the $C$-term includes the contributions of slow mass transfer in the stationary and mobile phases. This plot is similar to that in Figure 1.17 for flow through an open column coated with a stationary phase, except that here, with a packed column, the $A$-term is present and makes a constant contribution at all velocities. As in Figure 1.17, the $B$-term dominates at low average linear velocities, and the $C$-term dominates at high average velocities. The arrow indicates the optimum average linear velocity ($\bar{u}_{opt}$), meaning the velocity that produces the lowest plate height and therefore the narrowest peaks. As noted in the text, however, we frequently operate at linear velocities higher than the optimum, accepting the slightly broader peaks that result in order to decrease the total analysis time.

### 1.2.7. Practical Consequences of Broadening Theory

Having looked at the physical processes represented by each term, it is possible to combine the $A$-, $B$-, and $C$-terms and plot broadening as a function of linear velocity. Figure 1.22 shows the dependence of each individual term in the van Deemter equation on flow rate and the result when all of the terms are combined. From this plot, it is clear that the $B$-term – broadening due to longitudinal diffusion – makes the largest contributions to $H$ at low linear velocities. At higher linear velocities, the $C$-term dominates in regard to the contribution they make to broadening.

The lowest point on the curve corresponds to the linear velocity (and hence the experimental flow rate) at which $H$ is minimized, meaning the flow rate that generates the highest efficiency, $N$, possible for the system and conditions being studied. While this is the optimum flow rate in terms of broadening effects, it is common to operate at flow rates above the optimum, because higher flow rates mean faster analyses and more samples analyzed in a given period of time (e.g., more samples per day). For this reason, current emphasis in chromatography is being placed on minimizing the $A$- and $C$-term contributions to broadening. By keeping the slope in the high linear velocity region of the curve low, the flow rate can be increased significantly while incurring an acceptable increase in broadening, as demonstrated in Figure 1.23, curve "$a$." If the slope at high velocities is steep, as it is in curve "$c$," then increasing the flow rates leads to dramatic increases in $H$ and much broader peaks,

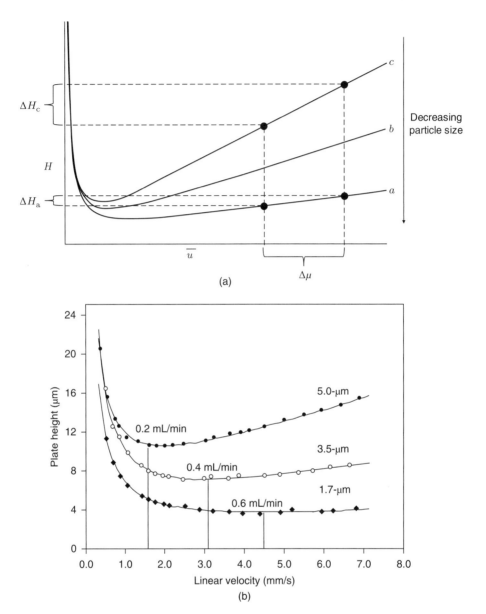

**FIGURE 1.23** (a) Depiction of the effect of decreasing particle size on plate height, in which the curve labeled (c) corresponds to larger particles and (a) corresponds to smaller particles. The vertical lines show that with smaller particles, a given increase in the linear velocity ($\Delta\mu$) produces a much smaller corresponding increase in plate height ($\Delta H$) compared to larger particles. This means that smaller particles can be used at higher velocities without significantly increasing band broadening. (b) Experimentally determined van Deemter curves for 5, 3.5, and 1.7 μm particles. The vertical lines correspond to the optimum linear velocity for each column (i.e., the velocity that produces the smallest plate height and thus the narrowest peak). Again we see that the C-term in the high linear velocity region is the least steep for the smallest particles, allowing for higher velocities with minimal additional band broadening. Conditions: 2.1 mm inner diameter column; 50/50 v/v (ACN/water); 210 nm detection; 45°C, benzylphenone as analyte; ultra-high performance LC system. (*Source:* Reprinted with permission from *Amer. Pharm. Rev.* 2008, 11, 24–33, from the authors of that article, and from Dr. Michael Dong who modified the original image for publication in *LC/GC North Am.*, 2014, 32, 553–557, which is the version shown here.)

which can lead to peak overlap, ruin the separation, and make quantitation of solutes less reliable or impossible. These considerations drive much of the practical work being done in chromatography today, including the manufacturing of smaller particles and particles that minimize broadening contributions arising from slow mass transfer. Figure 1.23b shows the dramatic difference in plate height obtained with particles of different diameter. The 1.7 μm particles produce significantly smaller plate heights than either the 3.5 or 5 μm particles.

Current work is also aimed at other instrumental design aspects that minimize broadening that occurs when the molecules are *outside* of the column, such as broadening due to the injection process and broadening that occurs in the tubes that transfer the solutes from the end of the column to the detector (so-called extra-column broadening effects).

Careful consideration of all of these factors has led to significant advances in the speed of LC and GC separations, making it possible to analyze complex samples containing tens or hundreds of compounds in just seconds or minutes.

## 1.3.  GENERAL RESOLUTION EQUATION

In the previous sections, we have seen all of the factors that affect the ability to separate two components, A and B, from one another, including overall retention ($k$), separation factor ($\alpha$), and band broadening as measured by the number of theoretical plates ($N$). These factors can be combined in an overall equation known as the general resolution equation:

$$R = \frac{\sqrt{N}}{4}\left(\frac{\alpha - 1}{\alpha}\right)\left(\frac{k_B}{1 + k_{ave}}\right) \tag{1.35}$$

where $k_B$ is the retention factor of the more retained solute of interest and $k_{ave}$ is the average retention factor for the two compounds whose resolution is being measured. In practice, the resolution between two peaks is only of interest when the two peaks elute near one another – in other words, when $k_A \approx k_B$ such that $k_B \approx k_{ave}$. Given this, for simplicity, $k_B$ or $k_{ave}$ can be used in both the numerator and denominator in the third term in Equation 1.35. An $R$ value of 1.5 typically produces peaks that are just baseline resolved for peaks of approximately the same height.

From Equation 1.35, it can be concluded that high $N$ values (narrow peaks), high $\alpha$ values (large differential retention), and high $k$ values (long retention) increase resolution. Figures 1.24–1.26 show the dependence of resolution on each of these three variables in this equation. Several practical conclusions can be reached by analyzing these graphs:

1. Because of the square root dependence of $R$ on $N$, increasing $N$, if possible, reaches a point of diminishing returns (Figure 1.24). Attention to the band-broadening details discussed above can contribute to increasing $N$ by decreasing band broadening, but $N$ is largely determined by the quality and characteristics of the column, including particle diameter, film thickness, and packing quality. So in practice, improvements in $N$ that are large enough to influence $R$ often cannot be made without investing in better columns or instrumentation.

2. Resolution can be improved by increasing retention ($k$) when retention is low, but at higher retention factors, this effect quickly levels off (Figure 1.25). In fact, the third

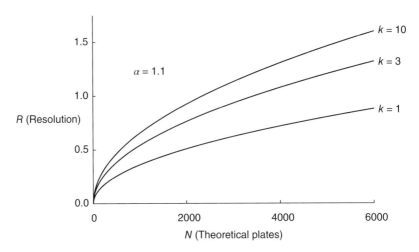

**FIGURE 1.24** Resolution ($R$) as a function of the number of theoretical plates ($N$) on a column for three different values of the retention factor, using the general resolution equation (Equation 1.35). The separation factor, $\alpha$, is set at 1.1. Note that $R$ rapidly increases as $N$ increases at low plate numbers, but $R$ increases more slowly at higher $N$. The square root dependence of $R$ on $N$ mitigates the effect on $R$ of increasing $N$ as $N$ approaches high values.

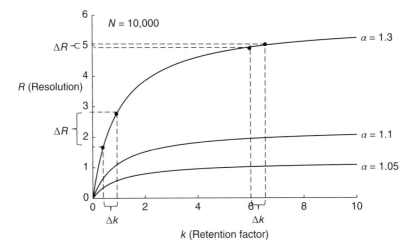

**FIGURE 1.25** Effect of retention factor ($k$) on resolution ($R$) at three different separation factors ($\alpha$), calculated using the general resolution equation (Equation 1.35) with $N = 10{,}000$. As illustrated by the dashed lines, increasing the retention factor ($\Delta k$) when the retention factor is low increases $R$ ($\Delta R$) considerably. When retention is already high ($k = 6$–10), increasing retention has little effect on improving resolution, but it makes the analyses take longer. It is also clear from this graph that small increases in the separation factor even from 1.05 to 1.10, improve resolution substantially at every retention factor. When viewing this plot, recall that a resolution of 1.5 is generally associated with baseline resolution for peaks of the same height. Values under 1.5 suggest overlapped peaks.

term in Equation 1.35 approaches a maximum value of 1 at high $k$. In addition, the way to increase $k$ is to increase the retention times of the solutes, thus making the analysis times longer, resulting in fewer analyses per day.

3. Small changes in separation factor ($\alpha$), particularly near $\alpha \approx 1$ (perfectly overlapped peaks) make large improvements in resolution (Figure 1.26). Thus, by changing the relative retention (i.e., the extent to which one solute is retained relative to another) by a small amount, large gains in resolution can be made. *Thus, in practice, the most*

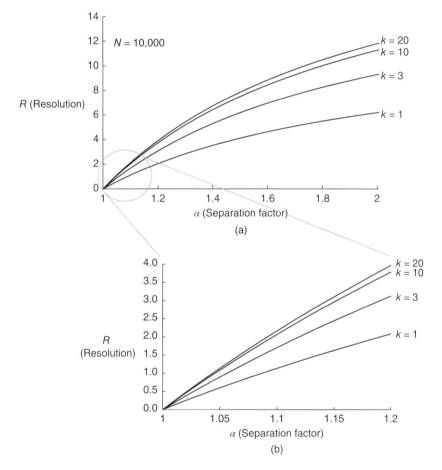

**FIGURE 1.26** Effect of separation factor ($\alpha$) on resolution ($R$), using the general resolution equation with $N = 10,000$ at varying values of retention factor ($k$). (a) This plot shows that as separation factors increase, so does resolution. However, the effect begins to level off at high separation factors. The plot also reinforces that there is little gain in resolution once retention factors exceed $k = 10$, but significant increases in analysis time result. The inert (b) shows an expanded view of the circled region of plot A. It is clear that small improvements in the separation factor can lead to relatively large improvements in resolution when the separation factor is initially low (i.e., between 1.0 and 1.2). Separation factors are often changed by varying the mobile phase composition (common in LC) or the temperature (common in GC). Recall, baseline resolution is generally associated with $R = 1.5$. These plots show that even small values of the separation factor can achieve baseline resolution of peaks if the solute retention factors are high enough and $N$ is large enough.

*important variable in the general resolution equation (assuming adequate N and k) is the separation factor.* The operating parameters that affect separation factors in GC and LC are discussed in detail in their respective chapters.

The theory underpinning the general resolution equation provides guidance for changing important operating variables to improve resolution. It drives the way we conduct chromatography. To measure resolution, however, we use data taken from the chromatogram and the equation

$$R = \frac{t_{r,B} - t_{r,A}}{\frac{1}{2}(W_A + W_B)} \tag{1.36}$$

where compound B elutes later than compound A and $W$ is the baseline width of the peaks determined graphically as shown earlier in Figure 1.7.

Again in this equation, we see that bigger differences in retention time (i.e., large $t_{r,B} - t_{r,A}$) and narrow peaks (small $W$'s) lead to higher resolutions, just as we saw in the general resolution equation. The two equations complement each other in that Equation 1.36 provides a way *to measure* resolution while the general resolution equation shows which variables to change *to improve* it.

---

**EXAMPLE** *1.7*

---

If *p*-xylene and *m*-xylene elute at 11.38 and 11.53 min, respectively, with a dead time of 3.23 min, on a column with 124,000 theoretical plates, what is the resolution of the peaks? Are the peaks baseline resolved? Assume the peaks are the same height.

**Answer:**

$$k_{p\text{-xylene}} = \frac{11.38\,\text{min} - 3.23\,\text{min}}{3.23\,\text{min}} = 2.523$$

$$k_{m\text{-xylene}} = \frac{11.53\,\text{min} - 3.23\,\text{min}}{3.23\,\text{min}} = 2.570$$

$$\alpha = \frac{k_{m\text{-xylene}}}{k_{p\text{-xylene}}} = \frac{2.570}{2.523} = 1.019$$

$$R = \frac{\sqrt{124{,}000}}{4}\left(\frac{1.019 - 1}{1.019}\right)\left(\frac{2.570}{1 + 2.547}\right)$$

$$R = 1.189 \approx 1.19$$

Baseline resolution is associated with $R = 1.50$, so these peaks are not baseline resolved and therefore overlap to some extent.

**Another question:**
How many plates are necessary to achieve baseline resolution (i.e., $R = 1.50$) for these compounds assuming retention times and the dead time do not change?

**Answer:**
$N = 197{,}000$ plates.

## 1.4.  PEAK SYMMETRY

For a variety of reasons, symmetric peaks are generally favored over asymmetric peaks in chromatography. Examples of asymmetric peaks are shown in Figure 1.27. Peaks that rise gradually but then fall off sharply are referred to as being "fronted," whereas peaks that rise quickly but trail off more slowly over time are referred to as "tailed."

Asymmetric peaks generally indicate that some aspect of the analysis is not optimal. This can arise from a range of phenomena that are beyond the scope of this text. If observed in practice, the source(s) of the asymmetry should be identified and corrected if possible.

The degree of peak asymmetry is measured by the peak asymmetry factor (AF), defined as the ratio of the peak $1/2$-widths at a given peak height, often taken at 10% of the total height, leading to

$$AF = \frac{b}{a} \tag{1.37}$$

where "$a$" and "$b$" are defined in Figure 1.27.

---

**EXAMPLE** *1.8*

Estimate the asymmetry factor for peak (*b*) in Figure 1.27.

**Answer:**

$$AF = \frac{b}{a} \approx \frac{8.040 - 8.028}{8.028 - 7.978} \approx 0.24$$

**Another Question:**
Estimate the asymmetry factor for peak (*c*) in Figure 1.27.

**Answer:**

$$AF = \frac{b}{a} \approx \frac{8.024 - 7.972}{7.972 - 7.960} \approx 4.3$$

Note: Answers will vary depending on how the times are estimated. It is important, however, to recognize that the times should be estimated from the position of the curve at 10% of the maximum height (not at the baseline) as shown in Figure 1.27.

---

## 1.5.  KEY OPERATING VARIABLES

From the theory above, it is evident that some of the key variables that must be controlled in chromatography include

1. retention;
2. separation factor (i.e., selectivity);
3. flow rate; and

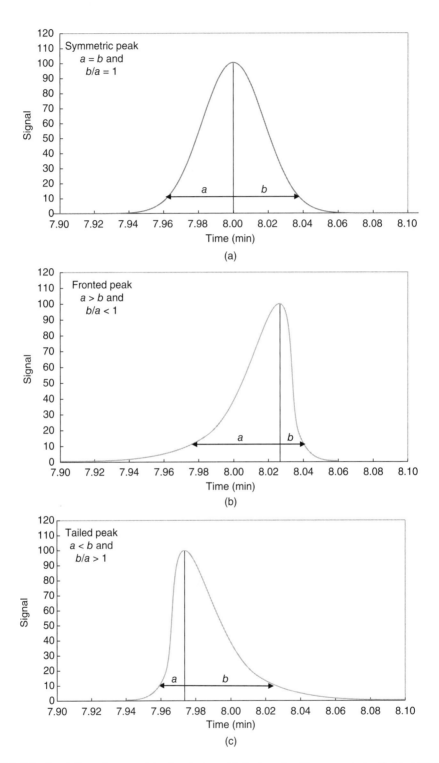

**FIGURE 1.27**　Depiction of symmetric, fronted, and tailed peaks. The asymmetry factor (AF) is measured as the ratio of $b/a$, where $b$ and $a$ are determined as shown in the figure at 10% of the peak height. Asymmetric peaks often indicate that something is not optimal and the source of the asymmetry should be investigated.

**4.** factors that affect broadening (particle size, stationary phase thickness, etc.)

The retention and separation factor in GC analyses are generally altered by changing the *temperature* at which the separation is done. The nature of the stationary phase impacts the fundamental intermolecular interactions that cause retention.

In liquid chromatography, the *composition of the liquid mobile phase* plays an important role in the retention and separation factor of solutes. The stationary phase material and the nature of the particles are also key variables in controlling LC separations.

These variables and specifically how they are adjusted in GC and LC are described in detail in the following chapters.

## 1.6. INSTRUMENTATION

While both GC and LC are fundamentally based on solute partitioning, the instrumentation used in the techniques is different due to the different demands arising from pumping gases versus liquids through columns. The methods of detection are also quite different. Because of these differences, the instrumentation used in each technique warrants its own discussion. Therefore, just like the key operating variables described above, the specific issues of instrumentation are described in more detail in the chapters that deal with GC and LC specifically.

## 1.7. PRACTICE OF THE TECHNIQUE

In the following two sections, we focus on one of the main reasons to use chromatography, namely to quantify the concentrations of various components in a mixture. As with many analytical techniques, this requires the construction of a calibration curve and often benefits from the use of internal standards or the method of standard additions to improve the accuracy and precision of the results.

### 1.7.1. Quantitation

A primary use of chromatography is the quantitation of individual components in a mixture, such as the amount of dextromethorphan (a cough suppressant) in a liquid cold medication or the concentration of an organic pollutant in a lake. While specific detection methods and their pros and cons are discussed in the GC and LC chapters, we note here that peak heights and peak areas, regardless of detector type, typically correlate with analyte concentration. Peak areas are generally favored for quantitative purposes because they produce better precision. However, in some cases, peak heights may be preferable. Based on the correlation of peak area with concentration, calibration curves are constructed by injecting solutions with known concentration of the analyte(s), measuring peak areas or heights, and plotting signal versus concentration as is typically done for calibration curves (see Figure 1.28). The sample, containing the analyte of unknown concentration is then injected under the same conditions, and the slope and intercept of the best fit line through the calibration data is used to calculate the concentration of the analyte in the sample.

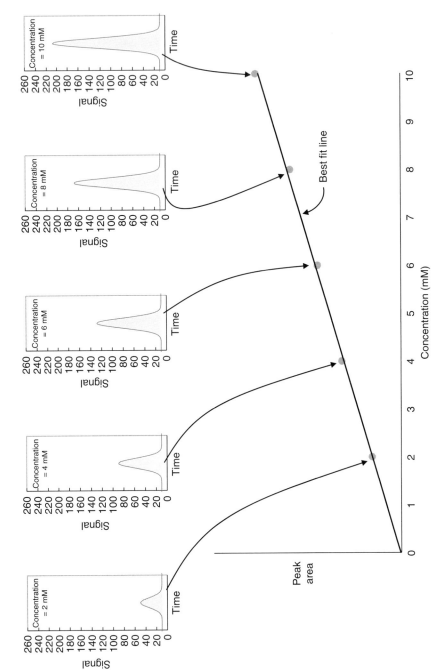

**FIGURE 1.28** Example of a calibration curve based on chromatographic peak area of the analyte.

Peak areas are generally obtained via computer software that computes the area under the curve, using peak recognition parameters that determine where a peak starts and ends. These parameters are set by the user and it is important for the user to visually inspect the chromatograms to ensure that these parameters lead to satisfactory recognition and integration of peaks.

### 1.7.2. Internal Standards and the Method of Standard Additions

Internal standards are frequently used in chromatography to compensate for variability arising from the injection process. As is always the case, the internal standards in chromatography are selected so that they generally mimic the behavior of the analyte(s). This means they should have comparable retention (although not identical as this would cause peak overlap) and generally produce similar detector responses as the analyte(s). In cases where the sample matrix may affect the response of the analyte, the method of standard additions is used to improve the accuracy of analyses.

## 1.8. EMERGING TRENDS AND APPLICATIONS

As hinted at in some of the earlier sections, one of the trends in chromatography is toward faster separations, driven by improvements in particle manufacturing and instrument design – particularly in the ability to create systems that can pump liquids at high pressures that are created by columns packed with small particles. In some ways, it is misleading to give the impression that the quest for faster separations is a recent one. Theory has driven practice in this regard for decades, and chromatographers continue to push the boundaries as new columns and instruments become available.

Another area that has received much consideration recently is the use of coupled columns that have significantly different stationary phases that maximize separation factors (i.e., selectivity).

Both of these areas, other work currently being pursued in both GC and LC, and applications of the techniques are described in greater detail in their respective chapters.

## 1.9. SUMMARY

Many areas of science, including biochemistry, biotechnology, chemical manufacturing, forensic science, environmental science, and the pharmaceutical industry, use chromatography to analyze samples such as blood, air, water, gasoline, food, medications, and crime scene evidence.

This chapter examined the fundamental concepts of partitioning, retention, separation, and resolution that are common to all modes of chromatography. In the following chapters, we describe how these concepts are put into practice in GC and LC specifically. We also look at how instruments are designed to introduce samples into the column, how mobile phase is pumped through the column, how the chemical composition of the stationary and mobile phases are manipulated to control retention, separation factors, and resolution, and how detectors are designed to sense the presence, concentration, and in some cases, the identity of the molecules as they elute from the column. There's a lot to look forward to!

## PROBLEMS

**1.1** (a) The water-to-hexadecane distribution coefficients for valeronitrile and octanal are 3.236 and 462.4, respectively. Which partitions out of water into hexadecane more favorably? Rationalize this based on the solute structures using arguments related to intermolecular interactions.

(b) Suppose 1,000,000 molecules of each are placed in a container with 500.0 mL of water and 500.0 mL of hexadecane. The container is shaken and allowed to come to equilibrium. How many molecules of valeronitrile and octanal are in the water phase at equilibrium? How many of each are in the stationary phase? Note that here again we are modeling a reversed-phase separation where water is the mobile phase and the hexadecane represents a less polar, organic-rich environment.

(c) Suppose that after equilibrium is achieved, the 500.0 mL of aqueous phase from this system are drawn off and added to a second container that has 500.0 mL of fresh, pure hexadecane in it already. This new system is shaken and allowed to come to equilibrium. How many molecules of valeronitrile and octanal are in the water phase at equilibrium? How many of each are in the stationary phase?

(d) What has happened to the mixture of valeronitrile and octanal that at the beginning had an equal number of both molecules in it?

(e) What would happen if this process of moving the mobile phase onto fresh stationary phase were repeated multiple times?

(f) Repeat the calculation in part (b), but this time let the aqueous phase be 950.0 mL and the hexadecane phase be just 50.0 mL. How does this affect the distribution of the solutes?

**1.2** (a) Calculate the retention factors for dichlorvos and mevinphos (two pesticides) that elute at $t = 22.84$ and $31.53$ s, respectively, with a dead time of 3.80 s in a fast GC separation.

(b) Based on these retention factors, for every one molecule of each solute in the mobile phase at any point in time, how many are in the stationary phase?

(c) Which solute spends more time in the stationary phase?

(d) Which solute spends more time in the mobile phase? Hint: think carefully about this.

**1.3** Calculate the phase ratio for a 30.0-m column with 0.530-mm inner diameter and 5.00-μm film thickness. It is quite helpful to sketch a picture to visualize the different volumes involved and how to calculate them.

**1.4** Which would increase retention for all solutes in a GC separation – using a column with a thinner film or a thicker one while keeping the inner diameter constant?

**1.5** (a) Calculate the separation factor given the following data:

$t_m = 1.56$ min

$t_{bromobenzene} = 3.87$ min

$t_{iodobenzene} = 4.57$ min

(b) If conditions were changed such that (1) the dead time remained the same, (2) the retention time of iodobenzene decreased more than that of bromobenzene, and (3) iodobenzene still eluted after bromobenzene, would the separation factor increase or decrease?

1.6 Calculate the HETP for a 25.0-cm RPLC column based on a peak that elutes at 17.456 min with a peak width at half maximum of 0.276 min.

1.7 Which process, longitudinal diffusion or slow mass transfer in the mobile phase, contributes more significantly to band broadening at high mobile phase velocities? Which dominates at low velocities?

1.8 Name and explain the two mechanisms that reduce band broadening arising from multiple flow paths through a packed bed.

1.9 Assuming all else is equal, which parameter in each of the following pairs would you choose in order to decrease broadening? Assume that the average linear velocity being used is above the optimum in each set.

(a) $N_2$ versus $H_2$ as the mobile phase in GC.
(b) $d_p = 2.0 \, \mu m$ versus $5.0 \, \mu m$ particles in RPLC.
(c) $0.25 \, \mu m$ versus $1.00 \, \mu m$ film thickness in GC.
(d) Increase the linear velocity versus decrease the linear velocity in either GC or RPLC (recall that the problem states that the velocity is already above the optimum velocity).

1.10 (a) Use the following conditions for octane as a solute, diffusing in He gas as the mobile phase, in a wall-coated capillary GC column, to calculate $H$ at the following average velocities: 2, 5, 10, 15, 20, 25, 30, 35, 40, 50, 60, 70, 80, 90, and 100 cm/s.

| Variable | Value |
|---|---|
| $d_f$ | $0.25 \, \mu m$ |
| $D_s$ | $1.20 \times 10^{-5} \, cm^2/s$ |
| $D_m$ | $0.191 \, cm^2/s$. This is the diffusion coefficient of octane in He, which is frequently used as the mobile phase in GC |
| $k$ | 3.25 |
| i.d. (inner diameter) | 0.250 mm |
| Length | 30.0 m |

It is helpful to use a spreadsheet and *calculate each of the three terms in the relevant equation explicitly* so you see how they each vary as a function of linear velocity. Be careful with units.

(b) Make a plot of the contribution of each of the three terms versus linear velocity.
(c) Make a plot of $H$ versus linear velocity.
(d) Use the plot to approximate the optimum linear velocity for this given set of conditions.
(e) Using the same spreadsheet, substitute the $D_m$ for octane in $N_2$ ($D_m = 0.046 \, cm^2/s$). Compare the two plots of $H$ versus linear velocity and compare the optimum velocities found for both gases.

**1.11** Make a plot of $H$ versus $k$ by calculating $H$ at $k = 0.1, 0.25, 0.50, 0.75, 1.0, 1.5, 2, 3, 4, 5, 6, 7, 8, 9, 10$ using Equation 1.23 and the following parameters (be careful with units).

| Parameter | Value |
|---|---|
| $\gamma$ | 0.700 |
| $D_m$ | $1.00 \times 10^{-5}$ cm$^2$/s |
| $\bar{u}$ | 0.300 cm/s |
| $\lambda$ | 0.500 |
| $d_p$ | 2.00 μm |
| $\omega$ | 0.500 |
| $d$ | 2.00 nm |
| $D_s$ | $3.00 \times 10^{-6}$ cm$^2$/s |
| $q$ | 0.133 |

**1.12** If a peak has an asymmetry factor of 0.73, is it tailed or fronted?

**1.13** (a) Calculate the resolution for the following two antihistamines separated by RPLC. Assume equal peak heights and Gaussian peak shapes.

| | $t_r$ (min) | $W_{baseline}$ (min) |
|---|---|---|
| Imipramine | 6.912 | 0.095 |
| Amitriptyline | 7.028 | 0.099 |

(b) If the dead time of this separation is 1.44 min, using the general resolution equation, how many theoretical plates does the column have?

(c) If the column is a 10.0 cm RPLC column, what is the plate height?

**1.14** The same solute was analyzed on two GC columns and yielded the following data:

| | $t_r$ (min) | $W_{1/2}$ (min) |
|---|---|---|
| Column 1 | 10.20 | 0.072 |
| Column 2 | 24.965 | 0.138 |

(a) Which column produced the broader peak?

(b) Which column has more theoretical plates?

(c) Based on theoretical plates, which column is better (i.e., more efficient)?

(d) What does this problem illustrate about evaluating columns? Can evaluations be based on peak width alone?

**1.15** In GC we make the approximation that the contribution to zone broadening from solute diffusion longitudinally down the column while *in the stationary phase* is negligible compared to that arising due to longitudinal diffusion in the mobile phase.

(a) Why is this a good approximation when the mobile phase is a gas?

(b) Is the approximation equally valid if the mobile phase is a liquid instead of a gas? Explain why or why not.

**1.16** Diffusion constants for solutes depend on the viscosity of the medium in which they are diffusing. In that regard, does band broadening increase or decrease as the temperature of the column, and hence the mobile and stationary phase in it, is increased in a liquid chromatography system? Assume one is already operating at flow rates above the optimum.

## REFERENCES

1. Stephens, T.W.; Quay, A.N.; Chou, V.; Loera, M.; Shen, C.; Wilson, A.; Acree, W.E., Jr.; Abraham, M.H. *Global J. Phys. Chem.* 2012, *3*, 1–42.

2. Martin, A.J.P.; Synge, R.L.M. *Biochem. J.*, 1941, *35*, 1358–1368.

3. Kiran, E.; Levelt Sengers, J.M.H. (Eds.) *Supercritical Fluids: Fundamentals for Application*, Series E, Vol. *273*; Springer Science and Business Media: Dordrecht, 1994.

4. Taylor, G. *Proc. Roy. Soc. A* 1953, *219*, 186–203.

5. Taylor, G. *Proc. Roy. Soc. A* 1954, *223*, 446–468.

6. Taylor, G. *Proc. Roy. Soc. A* 1954, *225*, 473–477.

7. Aris, A. *Proc. Roy. Soc. A* 1956, *235*, 67–77.

8. Guichon, G.; Guillemin, C.L. *Quantitative Gas Chromatography for Laboratory Analyses and On-line Process Control*; Amsterdam: Elsevier Publishing Company, 1988. p. 93–110.

9. McGuffin, V.L. "Chromatography" in *Fundamentals and Applications of Chromatography and Related Differential Migration Methods – Part A: Fundamentals and Techniques*, 6th Ed.; Heftmann, E. (Ed.); Elsevier Science: Amsterdam, 2004.

10. Golay, M.J. in *Gas Chromatography*, E. Desty (Ed.); Butterworths Scientific Publications: London, 1958.

11. Knox, J.H.; Scott, H.P. *J. Chromatogr. A* 1983, *282*, 297–313.

12. Giddings, J.C. *Dynamics of Chromatography, Part I: Principles and Theory*; Marcel Dekker, Inc.: New York, 1965.

13. Giddings, J.C. *Unified Separation Science*; Hoboken: John Wiley and Sons, Inc., 1991, p. 262.

14. van Deemter, J.; Zuiderweg, F.; Klinkenberg, A. *Chem. Eng. Sci.* 1956, *5*, 271–286.

15. Desmet, G.; Broeckhoven, K.; De Smet, J.; Deridder, S.; Baron, G.V.; Gzil, F. *J. Chromatogr. A*, 2008, *1188*, 171–188.

16. Desmet, G.; Broeckhoven, K. *Anal. Chem.* 2008, *80*, 8076–8088.

## FURTHER READING

1. Cazes, J.; Scott, R.P.W. *Chromatography Theory*; Marcel Dekker, Inc.: New York, 2002.

2. Gidding, J.C. *Dynamics of Chromatography, Part 1, Practice and Theory*; Marcel Dekker, Inc.: 1965.

3. Giddings, J.C. *J. Chem. Educ.* 1967, *44*, 704–709.

4. Giddings, J.C. *Unified Separation Science*; John Wiley and Sons, Inc.: Hoboken, 1991.

5. Golay, M.J. in *Gas Chromatography*, Desty, E. (Ed.); Butterworths Scientific Publications: London, 1958.

6. Gritti, F. *LC/GC North Am.*, 2014, *32*, 928–940.

7. Horvath, C.; Lin, H.-J., *J. Chromatogr.* 1976, *126*, 401–420.

8. Horvath, C.; Lin, H.-J. *J. Chromatogr.* 1978, *149*, 43–73.

9. Huber, J.F.K. *J. Chromatogr. Sci.* 1969, *7*, 85–90.

10. Katz, E.; Ogan, K.L.; Scott, R.P.W. H. *J. Chromatogr.* 1983, *270*, 51–75.

11. Knox, J.H. *J. Chromatogr. A* 2002, *960*, 7–18.

12. Knox, J.H.; Parcher, J.F. *Anal. Chem.* 1969, *41*, 1599–1606.

13. Knox, J.H.; Scott, H.P., *J. Chromatogr.* 1983, *282*, 297–313.

14. Snyder, L.R.; Kirkland, J.J.; Dolan, J.W. *Introduction to Modern Liquid Chromatography*, 3rd Ed.; John Wiley & Sons, Inc.: Hoboken, 2010.

15. Wenzel, T. Separation Science – Chromatography Unit. http://www.bates.edu/chemistry/files/2012/01/wenzel-chrom-text-revised-12-113.pdf, last accessed March 15, 2015.

# GAS CHROMATOGRAPHY

# 2

Gas chromatography (GC) is an important technique used in industrial, academic, environmental, and government laboratories to analyze mixtures of volatile and semivolatile compounds. Its applications include quantifying components in complex mixtures, determining solvent purity, analyzing organic synthesis products, monitoring water and air quality, and detecting explosives. A specific application explaining how GC is used to determine the country of origin of an oil sample is discussed at the end of this chapter. This case study resulted from the first Persian Gulf War and had an impact on world events involving Russia, the United States, Iraq, Iran, and Dubai.

As an example of the analytical power of GC, consider the chromatogram shown in Figure 2.1 obtained by analyzing a sample of premium unleaded gasoline. In this figure, each peak represents a different compound. As you can see, gasoline contains well over 100 compounds! The peaks are distinct, meaning that the individual components have been separated by the gas chromatograph. In this figure, 31 of the compounds have been identified. The time at which a particular peak appears ($x$-axis) is dictated by the compound's chemical structure, and the size of each peak is related to the concentration of the component in the sample. Because gas chromatography yields important qualitative and quantitative information, it is necessary to understand this analytical technique and how to maximize its utility.

## 2.1. THEORY OF GAS CHROMATOGRAPHIC SEPARATIONS

The analysis of complex mixtures by gas chromatography relies on the components in the mixture partitioning between a gas phase and a stationary phase. More precisely, it relies on *differences* in the components' partitioning behavior. Figure 2.2 depicts this partitioning. As discussed in Chapter 1, the partitioning can also be written as an equilibrium process as shown here:

$$\text{Solute}_{(gas)} \rightleftharpoons \text{Solute}_{(stationary\ phase)}$$

with an equilibrium constant (also called the distribution constant or partition coefficient in chromatography), $K$, where

$$K = \frac{[\text{solute}]_{stat}}{[\text{solute}]_{gas}} \tag{2.1}$$

*Chromatography: Principles and Instrumentation*, First Edition. Mark F. Vitha.
© 2017 John Wiley & Sons, Inc. Published 2017 by John Wiley & Sons, Inc.

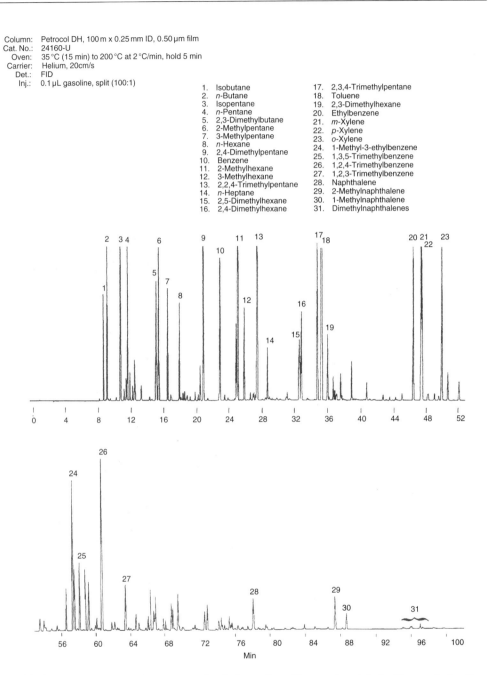

Column:    Petrocol DH, 100 m x 0.25 mm ID, 0.50 µm film
Cat. No.:  24160-U
Oven:      35 °C (15 min) to 200 °C at 2 °C/min, hold 5 min
Carrier:   Helium, 20cm/s
Det.:      FID
Inj.:      0.1 µL gasoline, split (100:1)

1.  Isobutane
2.  n-Butane
3.  Isopentane
4.  n-Pentane
5.  2,3-Dimethylbutane
6.  2-Methylpentane
7.  3-Methylpentane
8.  n-Hexane
9.  2,4-Dimethylpentane
10. Benzene
11. 2-Methylhexane
12. 3-Methylhexane
13. 2,2,4-Trimethylpentane
14. n-Heptane
15. 2,5-Dimethylhexane
16. 2,4-Dimethylhexane

17. 2,3,4-Trimethylpentane
18. Toluene
19. 2,3-Dimethylhexane
20. Ethylbenzene
21. m-Xylene
22. p-Xylene
23. o-Xylene
24. 1-Methyl-3-ethylbenzene
25. 1,3,5-Trimethylbenzene
26. 1,2,4-Trimethylbenzene
27. 1,2,3-Trimethylbenzene
28. Naphthalene
29. 2-Methylnaphthalene
30. 1-Methylnaphthalene
31. Dimethylnaphthalenes

**FIGURE 2.1**  Chromatogram of premium unleaded gasoline. (*Source:* Reprinted with permission from Supelco Inc, Bellefonte, PA, Petroleum/Chemical Application Guide, Bulletin 858D, 2000, and from R.L. Grob and E.F. Barry, *Modern Practice of Gas Chromatography*, 4th edition, 2004, page 696 with permission from John Wiley and Sons, Inc.)

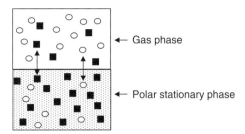

**FIGURE 2.2** Depiction of solute partitioning. Squares and circles represent solutes. In this figure, squares have a greater affinity for the stationary phase relative to the gas phase than do the circles.

The brackets indicate molar concentrations of the solute in the stationary phase (subscript "stat") and gas phase. In GC, the retention of compounds is dictated by solute intermolecular interactions with the stationary phase. Dispersion interactions tend to dominate in GC, with more specific interactions such as dipole–dipole and hydrogen-bonding interactions giving rise to additional retention. The strength of the interactions, and therefore the extent of partitioning of a solute between the gas and stationary phases, is strongly temperature dependent. *As the temperature of the system is increased, the gas-phase concentration of all solute molecules increases.*

In Figure 2.2, it is clear that the solutes have not been completely separated by a single partitioning. Clearly, something more must be done to achieve their separation. Nevertheless, this simple phenomenon of partitioning and its dependence on intermolecular interactions govern the separation. The task in understanding gas chromatography is to understand how the partitioning of solutes is controlled and how it can be used to separate all of the components in a mixture.

---

**EXAMPLE 2.1**

Determine the distribution constant, $K$, for a solute in a vessel containing 1.000 mL of stationary phase and 50.00 mL of gas phase, if at equilibrium $3.000 \times 10^{-3}$ moles of the solute are in the stationary phase and $24.00 \times 10^{-6}$ moles of solute are in the gas phase.

**Answer:**

$$K = \frac{[\text{solute}]_{\text{stat}}}{[\text{solute}]_{\text{gas}}} = \frac{\left( \dfrac{3.000 \times 10^{-3}\,\text{mol}}{1.000 \times 10^{-3}\,\text{L}} \right)}{\left( \dfrac{24.00 \times 10^{-6}\,\text{mol}}{50.00 \times 10^{-3}\,\text{L}} \right)} = 6250$$

Repeat the calculation if the quantities are 2.000 mL of stationary phase and 100.0 mL gas phase with $1.500 \times 10^{-3}$ moles of solute in the stationary phase and $78.00 \times 10^{-6}$ moles of solute in the gas phase.

**Answer:** $K = 962$.

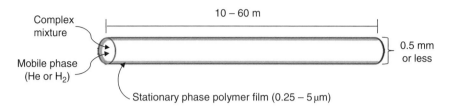

**FIGURE 2.3** Schematic of a wall-coated open tubular gas chromatography column.

### 2.1.1. GC Columns and Partitioning

Instead of establishing an equilibrium in a single container as depicted in Figure 2.2, gas chromatography is performed in long columns as depicted in Figure 2.3. In the most common mode of GC, the column is 0.5 mm or smaller in diameter, open in the center, and has a liquid-like polymeric stationary phase coated on the walls. This is referred to as open tubular or capillary GC. The sample is injected as a small, finite plug at one end of the column. A carrier gas such as $H_2$ or He serves as the mobile phase and, unlike the solute mixture, is *continuously* pumped through the column. The gas is chosen so as not to interact or react with the solutes in the gas phase and is thus chemically inert. It is there simply to push the solutes down the column. When a molecule partitions into the stationary phase, it does not move down the column. However, when the molecule leaves the stationary phase and enters the gas phase, it is swept down the column by the flowing carrier gas. By being swept to a new part of the column, the molecules are in contact with a new portion of stationary phase and reestablish an equilibrium between the stationary and mobile phases. Thus, the solutes reenter the stationary phase some distance down the column. In this way, the molecules can be thought of as tumbling or leapfrogging down the column. Solutes that are strongly attracted to the stationary phase spend a relatively long time in the stationary phase compared to the time they spend in the mobile phase. Conversely, compounds with weak interactions with the stationary phase spend relatively little time in the stationary phase. Therefore, they get swept to the end of the column in a shorter *total* amount of time than do solutes with stronger interactions with the stationary phase. In this way, a complete separation of components of a mixture can be achieved.

### 2.2. KEY OPERATING VARIABLES THAT CONTROL RETENTION

The goal in chromatographic separations is to obtain separation of the components of interest – in other words, to resolve the individual peaks. Physically, this means that the individual solute bands must move down the column at different speeds and elute at different times. In gas chromatography, the time it takes for a compound to travel from one end of the column to the other (i.e., its retention time, $t_r$) is controlled principally by three factors:

1. Temperature;
2. Mobile phase velocity; and
3. The chemical nature of the stationary phase (e.g., polar versus nonpolar).

To understand the practice of GC, it is necessary to understand how these variables affect retention and resolution. Each is discussed in the sections that follow.

### 2.2.1. Adjusting Retention Time: Temperature

The extent of interaction between the solutes and the stationary phase is adjusted by controlling the temperature of the system. The solutes gain energy as the temperature is increased. This provides a driving force for the solutes to more easily escape their intermolecular interactions with the stationary phase and enter the gas phase, resulting in decreased time spent in the stationary phase. Because the total time spent in the column is the sum of time spent in the stationary phase plus the time spent in the mobile phase ($t_r = t_s + t_m$), an increase in temperature results in a net decrease in retention time for all solutes. (Recall from Chapter 1 that the time spent in the mobile phase is the same for all species and is dictated by the gas flow rate, which is only very slightly affected by temperature.)

Consider two extreme scenarios. Imagine setting an extremely high column temperature. In this case, all the molecules will permanently reside in the gas phase and all will be swept through the column without being retained at all by the stationary phase. All of the different chemicals in the sample will elute at the same time and no separation would occur. From this, we learn that without retention there is no resolution, making any analysis impossible. In the other extreme, if the column is set at a very low temperature, all of the solutes will condense in the stationary phase at the very start of the column and never acquire enough thermal energy to leave it to enter the mobile phase. In this case, the solutes never make it to the end of the column to be detected. Again, no separation and no analysis occur.

Thus, the goal is to find an intermediate temperature in which solutes are retained yet still elute from the column in a reasonable amount of time. Furthermore, the temperature must be controlled such that solutes are not only retained but also separated from one another.

---

### EXAMPLE 2.2

Consider two closed containers each containing identical volumes of stationary and gas phase but held at different temperatures. The same solute is introduced into each container. In one case, the distribution constant of the solute is found to be 438 and in the other, it is found to be 53,754. Which one is held at the higher temperature? Explain your reasoning.

**Answer:**
$K = \frac{[solute]_{stat}}{[solute]_{gas}}$, so the lower $K$ indicates a lower stationary phase concentration of solute relative to the gas phase concentration. This occurs at higher temperatures because the higher temperature provides the solutes with enough energy to overcome intermolecular interactions in the condensed phase. So the lower distribution constant ($K = 438$) is associated with the higher temperature. Higher temperatures therefore lead to shorter retention times.

Gas chromatographs are usually operated between 40 and 400 °C, although subambient temperatures as low as −100 °C can be obtained on instruments using cryogenic cooling. The upper temperature limits the types of molecules that can be analyzed by GC because not all molecules are volatile in this range. For example, large polymers and most biomolecules such as proteins, RNA, and DNA cannot be analyzed by GC because they condense in the stationary phase, never enter the mobile phase, and thus do not elute from the column. By and large, most analytes migrate through the column at a reasonable speed when the column temperature is between 20 and 50 °C lower than the analyte's boiling point.

As we saw in Chapter 1, one of the most important parameters in chromatography is the separation factor, $\alpha$. The separation factor is a measure of the *relative* retention of two components. The greater the difference in retention times between two compounds, the greater the separation factor and the better the resolution, all else being equal. Because gas chromatography depends on the vaporization of solutes, the separation factor is related to the enthalpy of vaporization of the compounds of interest and the temperature at which the separation is conducted (Equation 2.2).

$$\ln \alpha = \ln \left( \frac{k_2}{k_1} \right) \approx \frac{\Delta \overline{H}_{vap,2} - \Delta \overline{H}_{vap,1}}{RT} + C \tag{2.2}$$

Here, $\Delta \overline{H}_{vap}$ is the molal enthalpy of vaporization of a pure liquid, $R$ is the universal gas constant, $T$ is the temperature of the system, and $C$ is a constant. The subscript "1" designates the earlier eluting species and "2" denotes the later eluting species.

It is important to note that this equation relies on the approximation that the types and strengths of intermolecular interactions that each solute experiences when it is in its pure bulk liquid form are identical, or nearly so, to the type and strength of intermolecular interactions it experiences when partitioned into the stationary phase. In other words, from the solute molecule's perspective, the equilibrium

$$\text{Solute}_{(pure\ liquid)} \rightleftharpoons \text{Solute}_{(gas)} \tag{2.3}$$

is identical to the equilibrium

$$\text{Solute}_{(stationary\ phase)} \rightleftharpoons \text{Solute}_{(gas)} \tag{2.4}$$

This approximation is most likely to hold when the solutes and the stationary phase are chemically similar (i.e., all are polar or all are nonpolar) such that the activity coefficients of the solutes in the stationary phase are close to unity or at least very similar to one another.

In these cases, the following approximation holds:

$$\Delta \overline{H}_{vap} = -\Delta \overline{H}_{soln} \tag{2.5}$$

where $\Delta \overline{H}_{vap}$ is the molal enthalpy of transfer of molecules from the pure compound to the gas phase (process shown in Equation 2.3) and $\Delta \overline{H}_{soln}$ is the molal enthalpy of the transfer of gas phase solute molecules into the stationary phase (the reverse of the process shown in Equation 2.4).

Accepting the approximations that lead to Equation 2.5, Equation 2.2 can be rewritten as

$$\ln \alpha \approx \frac{-(\Delta \overline{H}_{soln,2} - \Delta \overline{H}_{soln,1})}{RT} + C \qquad (2.6)$$

One important aspect of Equation 2.6 is that it shows that *when separating two different analytes, the bigger the difference in the strength of interactions they have with the stationary phase, the greater the separation factor.* This leads to better resolution between the two compounds. For example, the separation between hexane and nonane will be much greater than the separation of octane and nonane at the same temperature due to a bigger difference in dispersion interactions (and therefore enthalpies of solution) between hexane and nonane compared to the same difference between octane and nonane.

Perhaps even more important for the practice of gas chromatography, this equation also shows that lower temperatures generally favor larger separation factors for the separation of chemically similar solutes and thus generally provide better separations. However, lower temperatures also mean longer retention times and thus longer analyses. Conversely, higher temperatures result in shorter retention times but lead to decreases in the separation factor. *In practice, then, the ideal situation is to operate at the highest temperature that still provides separation of the compounds of interest.*

### 2.2.2. Adjusting Retention Time: Temperature Programming

In some mixtures, such as perfumes and gasoline samples, there are compounds that are quite volatile and therefore require relatively low temperatures for retention and separation, and other compounds that are large and have stronger intermolecular interactions with the stationary phase, requiring relatively high temperatures to elute. If these mixtures are injected at low temperatures to resolve the early-eluting compounds, the less volatile compounds may never elute or may come out in broad, unresolved peaks due to diffusion effects detailed in Chapter 1 (see the peak for compound 5 in Figure 2.4a). If the column temperature is held at a high temperature, the early-eluting, volatile compounds will not be separated (see Figure 2.4b).

In these cases, a technique known as temperature programming is employed (see Figure 2.4c). Very simply, temperature programming is the systematic variation of the column temperature while the analysis is conducted. For example, an analysis may begin by holding the column temperature at 40 °C for several minutes. This allows highly volatile components to be separated as they move down the column. After several minutes, the temperature is then increased by a certain amount each minute. This is referred to as a temperature ramp. Suppose the temperature ramp was 5°/min, and is employed for 20 min. The final temperature would then be 100°C *greater* than the initial temperature (e.g., 140 °C). It is typical to let the column remain at the high temperature for several minutes to allow any low-volatility components time to elute. It is also possible to program multiple ramps and holds into a single analysis.

Temperature ramps are usually anywhere from 2 to 20 °C/min, depending on the separation. The slope of the temperature ramp is chosen to optimize the separation of the

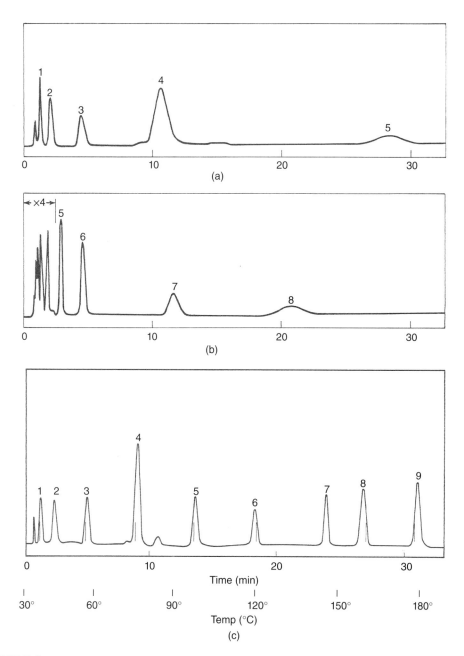

**FIGURE 2.4** The effect of temperature on chromatograms. (a) column at low temperature of 45 °C isothermal separation, (b) column at moderately high temperature 120 °C isothermal separation, and (c) temperature programmed separation from 30 to 180 °C with a rate of 4.8 °C. At 45 °C, the late-eluting peaks are severely broadened and all of the compounds have not yet eluted. At 120 °C, the early-eluting peaks overlap, which complicates quantitative analysis. With temperature programming, all of the peaks are eluted and resolved with narrow peak shapes. (*Source:* Reprinted with permission from W.E. Harris and H.W. Habgood, *Programmed Temperature Gas Chromatography*, John Wiley and Sons, Inc. 1966, page 10, also with permission from *Anal. Chem.*, 1960, 32, 450–453. Copyright 1960, American Chemical Society.)

components. In a complex mixture with solutes that span the entire range of volatility, slower temperature ramps lead to better resolution of all components. For a mixture of primarily low- and high-volatility compounds with little in between, steeper temperature gradients can be used in order to decrease the analysis time. We will see the employment of a temperature program for the analysis of crude oil samples in the application discussed at the end of the chapter.

While temperature programs are frequently necessary to achieve a separation of complex mixtures, it is often preferable to try to achieve the separation at a constant temperature – known as *isothermal analysis*. The reason for this is that when running replicate samples, if a temperature program has been used, the GC column must be returned from the final high temperature of the previous analysis to the initial low temperature for the start of the next analysis. This typically requires between 2 and 10 min. During this time, no analysis is being conducted. It is essentially wasted time. If the column is operated *isothermally*, no reequilibration time is required and the next analysis can begin immediately. In addition, the baseline in isothermal separations is typically flat, whereas with temperature programming, the baseline signal tends to increase as the temperature increases. This increase is due to fragments of the stationary phase that have decomposed, which elute from the column at elevated temperatures, and to highly retained solutes from previous analyses that have severely broadened as they have remained on the column through several analyses. The increasing baseline makes it more difficult to quantify the peak areas and heights, thus complicating the quantitation of the compounds. Of course, if temperature programs are the only way to achieve the desired separation, they must be used and the time and complications sacrificed for the sake of achieving the analysis.

### 2.2.3. Adjusting Retention Time: Mobile Phase Flow Rate

While temperature is the main variable used to obtain resolution, the mobile phase flow rate also impacts retention times and resolution. In gas chromatography, the mobile phase is an inert gas (i.e., nonreactive). Nitrogen, hydrogen, and helium are commonly used, with helium being the most common of the three. It is important to note, though, that changing the flow rate *does not* affect the distribution constants (i.e., $K$, also called the partition coefficient) of the solutes transferring between stationary and mobile phases.

Before discussing the effect of mobile phase flow rate on separations, the relationship between flow rate and linear velocity needs to be made explicit. The mobile phase volumetric flow rate, $F$, commonly measured in milliliters per minute, is related to its linear velocity through the column, $u$, measured in centimeters per second, as follows:

$$u\,(\text{cm/s}) = \frac{L}{t_{\text{m}}} \tag{2.7}$$

$$F\,(\text{mL/min}) = 60\pi r^2 u \tag{2.8}$$

where $L$ is the length of the column in centimeters, $r$ is the inner radius of the column in centimeters, and $t_{\text{m}}$ is the hold-up time, typically measured using an unretained species. The advantage of using linear velocity in the following discussion is that it, unlike the volumetric flow rate, is independent of the column diameter. However, as the equations show,

**TABLE 2.1    Linear Velocities Converted to Flow Rates as a Function of Column Diameter**

| | Flow rate in mL/min on 30 m column depending on column diameter | | | |
|---|---|---|---|---|
| Linear velocity (cm/s) | 0.53 mm | 0.32 mm | 0.25 mm | 0.10 mm |
| 20 | 2.65 | 0.97 | 0.59 | 0.09 |
| 40 | 5.29 | 1.93 | 1.18 | 0.19 |
| 80 | 10.59 | 3.86 | 2.36 | 0.38 |
| 100 | 13.24 | 4.83 | 2.95 | 0.47 |

flow rates and linear velocities can be interconverted if one knows the column diameter (see Table 2.1). In practice, the user enters a desired flow rate using the system software and then calculates linear velocities based on the column length and retention times.

---

**EXAMPLE 2.3**

An unretained solute takes 15.0 s to elute from a 30.0 m, 0.25 mm i.d. (inner diameter) column.

1. What is the linear velocity of the gas in centimeters per second?
2. What is the volumetric flow rate in milliliters per minute? Pay careful attention to units.

**Answer:**

1. linear velocity $= \frac{30.0\,m}{} \times \frac{100\,cm}{m} \times \frac{1}{15\,s} = 200\,cm/s$
2. volumetric flow rate $= (60.0\,s/min)(\pi)(0.0125\,cm)^2(200\,cm/s) = 5.89\,cm^3/min = 5.89\,mL/min$

Repeat the calculation if the unretained solute takes 12.0 s to elute from a 60.0 m, 0.18 mm i.d. column.

**Answer:**

Linear velocity $= 500\,cm/s$
Volumetric flow rate $= 7.63\,mL/min$

---

Figure 2.5 shows the relationship of plate height versus linear velocity in capillary GC. This graph contains a considerable amount of information of practical value. Increasing the mobile phase flow rate decreases analysis times because the solutes are swept through the column more quickly. Increasing flow rates, however, can also have a negative effect on resolution. As described in the sections about band broadening in Chapter 1, at high flow rates band broadening due to the parabolic flow in open tubes increases, causing an increase in plate heights as shown in Figure 2.5. This leads to lower resolution. The good news is that in GC, this increase in plate height at high velocities is small enough, separation factors are often large enough, and column plate counts are high enough, that analyses can be conducted at flow rates above the optimum value without a significant loss in resolution.

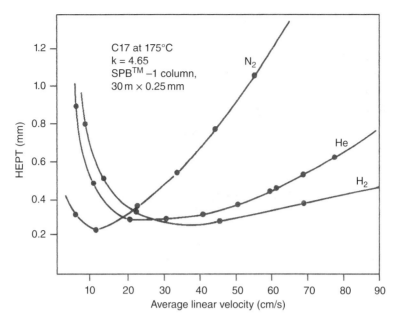

**FIGURE 2.5** Plate height (*H*) as a function of average linear velocity using $N_2$, He, and $H_2$ as the carrier gas. Nitrogen provides the lowest overall plate height possible, but at such a low linear velocity that it would result in very long analysis times. Hydrogen provides good plate heights and the least increase in plate height at high velocities (i.e., shorter analysis times). However, working with hydrogen can be dangerous. Therefore, helium, which offers performance that is comparable to that of hydrogen, is most commonly used as the carrier gas. (*Source:* Reprinted with permission from R.L. Grob and E.F. Barry, *Modern Practice of Gas Chromatography*, 4th edition, John Wiley and Sons, Inc. 2004, page 495.)

In the helium curve shown in Figure 2.5, we see that the optimum velocity (the velocity that provides the minimum plate height, H, and thus maximum resolution) is approximately 22 cm/s, with a corresponding plate height of approximately 0.28 mm. However, the slope of the curve at higher flow rates is not that steep, such that even at a linear velocity of 60 cm/s, the plate height is only 0.4 mm. For two compounds with arbitrarily chosen retention factors, *k*, of 8.0 and 8.4 analyzed on a 15 m column, the increase from a plate height of 0.28 mm at a linear velocity of 22 cm/s to a plate height of 0.4 mm at a velocity of 60 cm/s leads to a decrease in resolution from 2.46 to 2.06. Yet, the analysis time would go from approximately 11 min down to 4 min for the same change in linear velocity. Thus, the slight loss in resolution is well worth the significant decrease in retention time.

Figure 2.5 also shows that $H_2$ provides an even smaller increase in plate height, with increasing flow rates and a higher optimum flow rate in general. The potential of explosions with $H_2$, however, makes it a less attractive mobile phase gas than helium in cases where some resolution can be sacrificed for the sake of safety. Finally, nitrogen clearly is the least attractive alternative as it has the lowest optimum velocity and leads to significant increases in plate height at increased flow rates.

The difference in the plots between the three gases is due to their viscosities. Different viscosities mean that *solutes* diffuse in them at different rates. Nitrogen is the most viscous, leading to slower solute diffusion in $N_2$ than in He or $H_2$. Recall from Chapter 1 that solute diffusion in the radial direction is an important factor that helps reduce band broadening.

The slower diffusion in $N_2$ therefore causes a greater increase in the C-term in the Golay equation and therefore increased plate heights compared to He or $H_2$.

The take-home message in this section is that GC users must be aware of the approximate optimum flow rates of the gases they use in the event that the maximum resolution is required. However, the user must also be aware that small sacrifices in resolution made by increasing the flow rate may yield big benefits in decreased analysis time. The best selections of flow rates are made when both resolution and time of analysis are considered.

### 2.2.4. Adjusting Retention Time: The Column and the Stationary Phase

A variety of different types of GC columns exist. Some of the more commonly used columns and their important characteristics are listed in Table 2.2 and discussed in more detail below.

### 2.2.4.1.   *Open Tubular Capillary and Megabore Columns.* The vast majority of GC
analyses are conducted using capillary columns. They are narrow, open tubes typically made of fused silica, ranging from 10 to 60 m in length and 0.10 to 0.32 mm in inner diameter. The outside walls are coated with a brown polyimide polymer to give them structural support so that they can be coiled to fit inside the oven of a gas chromatograph. The inside walls are coated with the stationary phase of typical thicknesses ranging from 0.10 to 1 μm. A comparison of the diameter of the column ($10^{-3}$ m scale) to the thickness of the stationary phase ($10^{-6}$ m scale) reveals that the vast majority of the column volume is open space. Thus, these columns are often referred to as wall-coated *open tubular* (WCOT) columns. Columns with a 0.53 mm diameter are also commercially available. They are referred to as "megabore" columns. Megabore columns generally have thicker stationary phase films (1–5 μm) than conventional capillary columns and can therefore accommodate larger sample sizes (i.e., greater amount of sample injected into the column).

In WCOT columns, the stationary phase is a polysiloxane polymer that is covalently bonded to the silica walls. Thus, they are also referred to as bonded-phase columns. The chemical nature of the stationary phase can be varied to introduce different types and strengths of intermolecular interactions. Several of the most common stationary phase polymers are shown in Figure 2.6. Stationary phases such as the dimethyl polysiloxane and methyl phenyl polysiloxane are quite common and can interact with solutes through nonspecific intermolecular forces such as London-dispersion and dipole–induced dipole interactions. They therefore generally provide separations based on the bulk phase volatility (boiling points) of the solutes being separated.

Stationary phases such as cyanopropylphenyl and Carbowax introduce the potential for dipole–dipole and hydrogen-bonding interactions between solutes and the stationary phase. Thus, these phases can increase the retention of polar, hydrogen-bonding solutes such as hexanol and ethylamine, which may be only mildly retained by a dimethyl or methyl phenyl stationary phase. Additional polar phases that are not shown in Figure 2.6 include DEGS, TCEP, and FFAP (di(ethyleneglycol) succinate, 1,2,3-*tris*(2-cyanoehoxy)propane, and free fatty acid phase, respectively).

By judiciously selecting stationary phases, the relative retention of solutes in a mixture can be controlled. Separations that prove difficult on one phase may be more easily

**TABLE 2.2 Typical Characteristics for Common GC Columns**

| Column type | Characteristics | Length (m) | Inner diameter (mm) | Stationary phase thickness or weight percent | Characteristics |
|---|---|---|---|---|---|
| Capillary columns (wall-coated open tubular, WCOT) | Thin stationary phase layer bonded or physisorbed to a fused silica column | 5–60 | 0.10–0.32 | 0.10–1.0 μm | Very efficient (small plate heights). Require small sample volumes (less than 1 μL), otherwise easily overloaded leading to poor peak shapes |
| Megabore | Slightly wider inner diameters and thicker stationary phase layers than conventional capillary columns | 5–60 | 0.53 | 1.0–5.0 μm | Thicker stationary phases allow for larger sample volumes and therefore better for trace analyses |
| Porous-layer open tubular (PLOT) | A porous adsorbent surface forms a thin layer on the column walls. Solutes adsorb onto the surface as opposed to partitioning into a stationary phase | 15–30 | 0.32–0.53 | 3.0–20 μm | Increased surface area leads to more retention – good for gas analyses and larger sample sizes. Poorer plate heights than WCOT columns |
| Support-coated open tubular (SCOT) | Solid support particles coated with stationary phase form a thin layer on the column walls | 50 | 0.50 | Variable | Increased surface area leads to more retention – good for gas analyses and larger sample sizes. Poorer plate heights than WCOT columns. Largely replaced by PLOT columns |
| Packed columns | Wide diameter columns typically made of glass or metal, which are packed with small particles coated with a stationary phase polymer | 1.0–2.0 | 2.0 | 3–20% | Can handle much larger sample volumes than capillary phases due to increased amount of stationary phase. Good for gas analyses due to increased retention. Inferior plate heights and therefore more potential for overlapping peaks compared to capillary columns |
| Micropacked columns | Smaller diameters than conventional packed columns but larger than capillary columns, packed with stationary phase coated particles | 1.0–2.0 | 0.75–1.0 | 20% | Offer plate heights and sample columns between those of conventional capillary and packed columns |

**Rxi®-1ms,
Rxi®-1HT, Rtx®-1**
Dimethyl polysiloxane

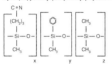

Similar to: (100%-methyl)-polysiloxane
Polarity: nonpolar
Uses: solvents, petroleum products,
[G1] pharmaceutical samples, waxes
[G2] [G38]

**Rxi®-5ms, Rxi®-5HT,
Rtx®-5, Rtx®-5MS**
Diphenyl dimethyl polysiloxane

Similar to: (5%-phenyl)-methylpolysiloxane
Polarity: slightly polar
Uses: flavors, environmental,
[G27] aromatic hydrocarbons
[G36]

**Rxi®-5Sil MS**
1,4-bis(dimethylsiloxy)phenylene
dimethyl polysiloxane

Similar to: (5%-phenyl)-methylpolysiloxane
Polarity: slightly polar
Uses: flavors, environmental, pesticides,
PCBs, aromatic hydrocarbons

**Rtx®-20**
Diphenyl dimethyl polysiloxane

Similar to: (20%-phenyl)-methylpolysiloxane
Polarity: slightly polar
Uses: volatile compounds, alcohols
[G28] [G32]

**Rtx®-1301, Rtx®-624,
Rtx®-G43**
Cyanopropylmethyl phenylmethyl polysiloxane

Similar to: (6%-cyanopropylphenyl)-
methylpolysiloxane
Polarity: intermediately polar
Uses: volatile compounds, insecticides
[G43]

**Rxi®-624Sil MS**

Similar to: (6%-cyanopropylphenyl)-
methylpolysiloxane
Polarity: intermediately polar
Uses: volatile compounds, insecticides, residue
solvents in pharmaceutical products

**Rtx®-35**
Diphenyl dimethyl polysiloxane

Similar to: (35%-phenyl)-methylpolysiloxane
Polarity: intermediately polar
Uses: pesticides, PCBs, amines,
[G42] nitrogen-containing herbicides

**Rxi®-35Sil MS**

Similar to: (35%-phenyl)-methylpolysiloxane
Polarity: intermediately polar
Uses: pesticides, PCBs, amines,
nitrogen-containing herbicides

**Rtx®-200**
Trifluoropropylmethyl polysiloxane

Similar to: (trifluoropropyl)-methylpolysiloxane
Polarity: selective for lone pair electrons
Uses: environmental, solvents, Freon® gases,
[G6] drugs, ketones, alcohols

**Rtx®-50**
Phenyl methyl polysiloxane

Similar to: (50%-phenyl)-methylpolysiloxane
Polarity: intermediately polar
Uses: FAMEs, carbohydrates
[G3]

**Rxi®-17**
Diphenyl dimethyl polysiloxane

Similar to: (50%-phenyl)-methylpolysiloxane
Polarity: intermediately polar
Uses: triglycerides, phthalate
[G3] esters, steroids, phenols

**Rxi®-17Sil MS**

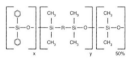

Similar to: (50%-phenyl)-methylpolysiloxane
Polarity: intermediately polar
Uses: triglycerides, phthalate
[G3] esters, steroids, phenols

**Rtx®-1701**
Cyanopropylmethyl phenylmethyl polysiloxane

Similar to: (14%-cyanopropylphenyl)-methylpolysiloxane
Polarity: intermediately polar
Uses: pesticides, PCBs, alcohols, oxygenates
[G46]

**Rtx®-65, Rtx®-65TG**
Diphenyl dimethyl polysiloxane

Similar to: (65%-phenyl)-methylpolysiloxane
Polarity: intermediately polar
Uses: triglycerides, rosin acids, free fatty acids
[G17]

**Rtx®-225**
Cyanopropylmethyl phenylmethyl
polysiloxane

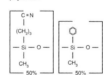

Similar to: (50%-cyanopropylmethyl)-
methylphenylpolysiloxane
Polarity: polar
Uses: FAMEs, carbohydrates
[G7] [G19]

**Rtx®-2330**
Biscyanopropyl cyanopropylphenyl polysiloxane

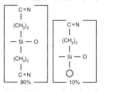

Similar to: (95%-cyanopropyl)-phenyl polysiloxane
Polarity: polar
Uses: cis/trans FAMEs, dioxin isomers, rosin acids
[G8] [G48]

**Stabilwax®, Rtx®-Wax,
Stabilwax®-MS**
Polyethylene glycol

Polarity: polar
Uses: FAMEs, flavors, acids, amines, solvents, xylene isomers
[G14] [G15] [G16] [G20] [G39]

**FIGURE 2.6** GC stationary phases from Restek Corporation. Other vendors offer similar stationary phase structures and use a similar numbering scheme, but replace the Restek prefixes (e.g., Rtx and Rxi) with letters specific to their companies. For example, J&W Scientific uses the letters "DB" and Supelco uses "SPB." (*Source:* Reproduced with permission of Restek Corporation.)

achieved on a different stationary phase. Note that in Figure 2.6, the columns are classified as nonpolar, slightly polar, intermediately polar, and polar. These labels refer to the type of molecules that are retained and separated by the columns and generally reflect the relative strength of intermolecular interactions on the columns. Thus, changing the chemical nature of the stationary phase changes the relative retention (i.e., separation factor) of solutes. In this way, the stationary phase can be used to adjust retention and the separation factor.

It should be noted, however, that labels such as "intermediate polarity" are vague and do not reveal information about the specific intermolecular forces such as dispersion, dipole–dipole, and hydrogen-bonding interactions that solutes experience on different columns. Retention and separation in chromatography ultimately rely on intermolecular interactions between solutes and the stationary and mobile phases. Changing the blend of interactions alters solute retention. In GC, this is achieved by changing the chemical structure of the stationary phase, which typically requires replacing one commercial column with an entirely different one. The choice of which column to choose as a replacement is guided by the chemical characteristics of the columns. These are measured using the Kovats Index and McReynolds constants. Describing these concepts at this point would lead us too far astray from our general discussion of GC columns. They are important concepts, however, and are explained in greater detail in Section 2.4.

---

## EXAMPLE 2.4

1. Classify each of the following phases as nonpolar, slightly polar, intermediately polar, or polar.
   (a) 20% diphenyl 80% dimethyl polysiloxane
   (b) 50% cyanopropylmethyl 50% methylphenyl polysiloxane
   (c) 6% cyanopropylphenyl 94% dimethyl polysiloxane
   (d) 35% diphenyl 65% dimethyl polysiloxane
2. Determine which is likely to produce the most retention for 2-pentanone, all else being equal.
3. What functional group on the phase you selected as you answer for question 2 is largely responsible for giving it its characteristic polarity?

**Answer:**
1. (a) Slightly polar
   (b) Polar
   (c) Intermediately polar
   (d) Intermediately polar.
2. All else being equal, 50% cyanopropylphenyl 50% methylphenyl polysiloxane would produce the greatest retention because 2-pentanone is a polar molecule and will interact strongly with the polar cyano groups via dipole–dipole interactions and also with the phenyl rings via dispersion and dipole–induced dipole interactions.
3. The polarity of the phase derives largely from the polar cyano function group.

In addition to selecting columns with different stationary phases, columns with different stationary phase thicknesses can also be selected. *Thicker stationary phases result in an increase in retention for all solutes relative to thinner stationary phases of the same chemical nature.* This can be understood by considering the fundamental retention process governing chromatography, namely, the partitioning of the solute between the gas mobile phase and the polymer stationary phase. Consider two 0.32-mm-diameter columns, one with a film thickness of 1 μm and the other with a 5 μm film thickness, as depicted in Figure 2.7. In *relative* terms, there has been a substantial increase in the *volume* of stationary phase ($V_s$) and a negligible decrease in the volume of the mobile phase ($V_m$, the open area) in going from a 1 to 5 μm film. Because the underlying intermolecular interactions causing partitioning have not changed, $K$ must be the same on the two columns. Now consider the distribution constant expression shown in Equation 2.9. In order to keep $K$ constant, as $V_s$ increases significantly and $V_m$ is essentially constant, the number of moles of solute in the stationary phase ($n_s$) must increase substantially to offset the increase in $V_s$ (i.e., as $V_s \uparrow$, the $n_s \uparrow$ and $n_m \downarrow$ to keep $K$ constant).

$$K = \frac{n_s/V_s}{n_m/V_m} = \frac{n_s V_m}{n_m V_s} = k\beta \qquad (2.9)$$

In this equation, $\beta$ is known as the phase ratio. We point out that we are following the IUPAC definition ($\beta = V_m/V_s$), but in other literature, the phase ratio is given the symbol $\phi$ and is defined as $V_s/V_m$ (i.e., the reciprocal form). Both are acceptable and the user simply has to be aware of which definition is being used. Because the total number of moles injected is independent of the stationary phase thickness, and because $n_{total} = n_s + n_m$, where $n_m$ represents the number of moles of solute in the mobile phase, as $n_s$ increases, $n_m$ must decrease. Thus, at any point in time, many more solute molecules reside in the stationary phase in thicker films than on thinner films. Because molecules do not move down the column when they are in the stationary phase, the increased number of molecules in thicker stationary phases means these phases provide increased retention relative to thinner phases. Thicker films are used to ensure sufficient retention and separation of highly volatile compounds. Conversely, thinner phases are used to achieve reasonable retention times, and hence reasonable analysis times, for lower volatility compounds.

From the above discussion, it is clear that both the *nature* of the stationary phase and its *thickness* affect *absolute* resolution. However, *only the nature of the stationary phase affects*

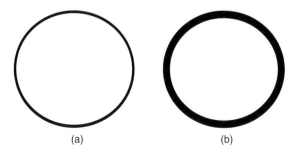

(a)                                    (b)

**FIGURE 2.7** Effect of film thickness on $V_s$ and $V_m$. The phase ratio, $V_m/V_s$, depends significantly on $V_s$ in wall-coated open tubular capillary chromatography because as $V_s$ changes a lot (e.g., doubles), $V_m$ does not show as significant a decrease.

*the separation* factor, $\alpha$, because the nature of the stationary phase, and not its thickness, dictates the solute equilibrium constants. This is shown in Equations 2.10 and 2.11 where B and A are two species being separated. Note that the phase ratio, $\beta$, which is a property of the column and therefore not solute dependent, cancels in Equation 2.11, and thus does not impact the separation factor.

$$k = \frac{K}{\beta} \tag{2.10}$$

$$\alpha = \frac{k_B}{k_A} = \frac{K_B/\beta}{K_A/\beta} = \frac{K_B}{K_A} \tag{2.11}$$

### 2.2.4.2. PLOT and SCOT Columns.

Two other classes of open tubular capillary columns are sometimes used in GC. In porous layer open tubular (PLOT) columns, a porous adsorbent such as alumina, a molecular sieve, or a polymer such as modified styrene/divinylbenzene, is coated onto the silica walls of an open tube. Solutes partition between the gas phase and the surface of the adsorbent and retention is governed by the solutes' gas–solid distribution constants. PLOT columns are used because the porous nature of the adsorbent considerably increases the surface area onto which solutes may adsorb, thus providing significant retention for highly volatile gaseous compounds, such as $O_2$, $N_2$, Ar, $H_2S$, $CS_2$, and $CO_2$, which are difficult to separate on standard WCOT columns.

In support-coated open tubular (SCOT) columns, solid support particles, typically made of diatomaceous earth, are coated onto the silica walls of an open tube. These support particles are then coated with a stationary phase liquid or polymer. While SCOT columns used to be common for the analysis of gases and volatile mixtures, they have largely been replaced by PLOT columns and microbore packed columns discussed below.

### 2.2.4.3. Packed Columns.

In contrast to open tubular GC, some GC columns are filled with small particles referred to as the packing material or solid support particles. These are often made of diatomaceous earth and are coated with a thin layer of stationary phase. Celite, Chromosorb P, and Chromosorb W are popular trade names for diatomaceous earth support particles. The particles range in size from 125 to 250 $\mu$m. The columns themselves often have significantly larger inner diameters than capillary columns, in the range of 2–4 mm, and are made of glass or metal. Because the particles resist the flow of carrier gas through the column (imagine blowing through an open straw compared to one packed with sand), packed bed columns are shorter than capillary columns to limit the pressures that must be applied in order to obtain reasonable retention times. Typical wide-bore packed columns are 2–5 m in length.

The advantage of packed columns is that the surface area is considerably greater than in capillary columns because each particle is coated with stationary phase – the same liquid phases discussed earlier such as Carbowax and methylsilicone (i.e., polydimethyl siloxane). This is analogous to a lake that contains hundreds of small islands having considerably more shoreline than the same size lake without islands. This increased stationary phase volume allows for larger sample volumes compared to capillary columns. On capillary columns, liquid sample volumes from 0.1 to 2 $\mu$L are common. Samples larger than this can overwhelm the stationary phase capacity, causing poor peak shapes and destroying resolution. On packed columns, however, much larger injection volumes can be used because of the much greater surface area and stationary phase volume. The greater stationary phase volume also significantly increases solute retention, which can be useful when analyzing highly volatile components. They are therefore useful for analyzing gases such as $CO_2$, CO, $CH_4$, $O_2$, and the rare gases.

The disadvantage of packed columns is their lower efficiency (i.e., smaller plate counts) compared to capillary columns. A typical 2 m packed column has approximately 2000–5000 plates, whereas a typical 30 m, 0.32 mm diameter capillary column has approximately 75,000–150,000 plates. Thus, resolution on packed columns is lower than on capillary columns, making the analysis of complex mixtures difficult on packed columns. This difference in plate number arises from the eddy dispersion caused by the presence of a packed bed, slower mass transfer due to increased volume of stationary phase, and shorter columns (2 versus 30 m).

***2.2.4.4. Micropacked Columns.*** Advances in the manufacture and packing of capillary columns have made it possible to combine the resolution afforded by capillary columns with the increased sample volume afforded by packed beds. These columns, referred to as micropacked columns, are 0.75–1.00 mm in diameter and are packed with 125–150 μm particles coated with a stationary phase. While not quite as efficient as open tubular capillary columns, packed capillary columns are considerably more efficient than wide-bore packed columns and offer greater sample capacity than open-tube capillary columns. Thus, larger samples can be injected and separated without the peak distortions that would occur in open-tube capillary columns and without the poor resolution often encountered in traditional packed columns.

As stated earlier, the different column types are summarized in Table 2.2.

### 2.2.5.   Adjusting Retention Time: Summary

In summary, absolute retention times in gas chromatography are influenced by manipulating

- the temperature of the system;
- the mobile phase flow rate;
- the chemical nature of the stationary phase;
- the type of the column used (open tube versus packed); and
- the volume of the stationary phase.

While temperature is the most commonly varied parameter, the practitioner of GC must nevertheless be aware of all of these variables and which are most likely to lead to improvements in specific separations. In other words, it is important to know which will not just alter retention, but also which should be changed in order to achieve the *separation factors* ($\alpha$) and *plate heights* ($H$) necessary for each particular sample to be analyzed.

### 2.2.6.   Measures of Retention

Most chromatographic software packages allow for the simple measurement of retention *times*. It is sometimes useful, however, to report retention in terms of the *volume* of mobile phase required to elute the compounds. In this case, the *corrected retention volumes* of a retained ($V_R^{\circ}$) and unretained ($V_M^{\circ}$) solute are given by

$$V_R^{\circ} = t_R \cdot F_{corr}^{\circ} \quad \text{and} \quad V_M^{\circ} = t_M \cdot F_{corr}^{\circ} \tag{2.12}$$

respectively, where $F_{corr}^{\circ}$ is the volumetric flow rate of carrier gas, typically given in milliliters per minute, corrected for temperature and pressure effects described as follows.

Two methods for determining flow rates are commonly employed in GC. The first relies on using a soap bubble meter or an electronic flow meter to measure the flow at the end of the column. A second method relies on measuring the retention time of an unretained solute and using column dimensions to calculate the volumetric flow from this measurement.

In modern instruments, the need for such measurements has been largely eliminated by the inclusion of instrumental controls and flow calibration techniques. Nevertheless, it is useful to know about these methods so that users can evaluate the reliability of the instrumental controls.

## USE OF SOAP BUBBLE METERS

In a soap bubble meter, a thin film of surfactant is carried upward within a cylindrical glass tube by the flow of carrier gas exiting from the column. The bubble meter, which looks like a buret, is graduated such that it is possible to observe the surfactant film rise from one mark to the next, where the marks represent volumes. To use a bubble meter, a stopwatch is started when the thin film crosses one mark, and stopped when it reaches some higher mark. The volume of gas represented by the rise in the film is divided by the time it took to deliver that volume, generating a flow rate in milliliters per minute ($F_m$ for "measured" flow).

It must be noted, however, that this represents the flow at the *outlet of the column* and as such it is being measured at a temperature that is different than the column temperature. Because gases are compressible and volumes are dependent on temperature, to obtain flow rates that represent the flow *inside the column*, the measured flow rate must by corrected for compressibility and temperature effects. In addition, because using a soap bubble meter means that water must be present, the carrier gas that is causing the bubble to rise becomes saturated with water. Thus, the pressure that the water exerts must be subtracted so that only carrier gas flow is being calculated.

To obtain the flow rate corrected for temperature, compressibility, and water vapor pressure, the following equation is used:

$$F_{corr}^{\circ} = F_m \cdot \frac{T_C}{T} \cdot \frac{P - P_{H_2O}}{P} \cdot j \tag{2.13}$$

where $F_m$ is the flow rate measured with the bubble meter, $T_C$ and $T$ are the temperature of the column (i.e., the oven temperature) and of the bubble meter (i.e., ambient temperature) in Kelvin, respectively, $P$ is the pressure at the bubble meter (atmospheric pressure), $P_{H_2O}$ is the vapor pressure of water at temperature $T$, and the last term, $j$, is called the compressibility factor and is calculated by

$$j = 1.5 \left( \frac{(P_i/P)^2 - 1}{(P_i/P)^3 - 1} \right) \tag{2.14}$$

where $P_i$ is the pressure of carrier gas at the inlet of the column, which is typically given by a pressure gauge of the instrument.

The compressibility factor is required because the flow of carrier gas through the column is established by applying a high pressure of the carrier gas to the front (injection side) of the column while the end of the column (detection side) is open to the atmosphere. Thus, a pressure drop exists across the column. Because gases are compressible, 1 mL of gas under the high pressure experienced at the column inlet expands as it travels through the column toward the low pressure outlet. Thus, the volumetric flow in milliliters per minute changes along the axis of the column and does so in a nonlinear way.

By employing these corrections, one can use Equation 2.12 to calculate the *corrected retention volume*, $V_R^\circ$, which represents the volume of mobile phase gas required to elute the compound at the average column pressure. Subtracting $V_M^\circ$ from $V_R^\circ$ yields the *net retention volume*, $V_N$ as defined in Equation 2.15

$$V_N = V_R^\circ - V_M^\circ \tag{2.15}$$

The *specific retention volume* $(V_g^\circ)$ at the column temperature and pressure averaged over column length is another parameter that is sometimes calculated. It is defined as

$$V_g^\circ = \frac{V_N}{w} \tag{2.16}$$

where $w$ is the mass of the stationary phase. Retention is greater on columns with more stationary phase because the solute partitioning equilibrium is shifted toward the stationary phase. $V_g^\circ$ normalizes for this and puts retention on a per gram of stationary phase basis. In the past, it was customary to multiply $V_g^\circ$ by the ratio $273.15/T_c$ to obtain the *specific retention time at* $0\,^\circ\text{C}$. The significance of this is that it allows separations conducted at different temperatures on different columns (that contain the same stationary phase) to be compared. An IUPAC report, however, recommends against this on the basis that doing so "will significantly distort the actual relationship between the retention volumes measured at different temperatures."[1-5] A general discussion of IUPAC recommendations for GC retention nomenclature and symbols can also be found in these references.

Regarding the thermodynamics of retention, in cases where the chemical nature of the solute and the stationary phase are similar (i.e., both are polar or both are nonpolar such that the solute activity coefficient is close to unity when dissolved in the stationary phase) $V_g^\circ$ can be related to the molal enthalpy of solution of the solute as shown in the following equation:

$$\ln V_g^\circ = \frac{-\Delta \overline{H}_s}{RT} + C' \tag{2.17}$$

where the constant, $C'$, is related to constant column parameters. Plots of $\ln V_g^\circ$ versus $1/T$ are linear and the slope yields $-\Delta \overline{H}_s/R$. Furthermore, because $V_g^\circ$ is ultimately a consequence of solute partitioning, it is related to the distribution constant, $K$, and therefore to the separation factor, $\alpha$. These relationships yield

$$\ln \alpha = \frac{-(\Delta \overline{H}_{s,2} - \Delta \overline{H}_{s,1})}{RT} + C \tag{2.18}$$

where 2 and 1 are two different solutes of interest being separated (where 2 elutes after 1) and C is a constant. As discussed, this equation is based on the approximation that the nature of the solutes and stationary phase are similar (i.e., all nonpolar or all polar). As highlighted earlier, this equation shows that the separation factor usually decreases with increasing column temperature because in most cases the later eluting peak has a higher molal heat of solution than the earlier eluting peak.

These calculations all essentially hinge on using a soap bubble meter to measure the flow of carrier gas exiting the column and subsequently calculating the average column flow. Another method of measuring column flow is based on the measured retention time of an unretained solute that has been detailed elsewhere.[6] This method offers the benefit that it does not require using a soap bubble meter, using instead the pressure gauge on most instruments, known column dimensions, and temperature and compressibility corrections to calculate the average flow in the column.

It is important to note that all of the above calculations rely on the column temperature being kept constant throughout the separation and thus do not rigorously apply when temperature programs are used. It is also important to note that many modern GC instruments have internal flow and pressure controls and can be operated in *constant pressure* or *constant flow* modes. In other words, many instruments have built-in features for calculating and controlling flows. Nevertheless, it is important to check the accuracy of the instrument. *To do so, the user must be aware of which mode is being used in order to have complete control and understanding of the analysis.*

## EXAMPLE 2.5

Calculate the net retention volume for a solute that elutes at $t = 5.43$ min under the following conditions: dimethyl polysiloxane column; column length, 30 m; column i.d. (inner diameter), 0.250 mm; column temperature, 150°C; ambient temperature, 22.0°C; inlet pressure, 15.0 psi; ambient pressure, 0.970 atm. An unretained species eluted at $t = 0.813$ min under the same conditions. The vapor pressure of water at 22.0°C is 0.383 psi. The column flow rate was measured using a soap bubble meter and found to be 1.99 mL/min.

**Answer:**
To find the corrected flow rate, the compressibility factor, $j$, is needed. The pressure units must be the same so the ambient pressure has been converted to psi.

$$j = 1.5 \left( \frac{(P_i/P)^2 - 1}{(P_i/P)^3 - 1} \right) = \frac{1.5[(15.0\,\text{psi}/14.26\,\text{psi})^2 - 1]}{[(15.0\,\text{psi}/14.26\,\text{psi})^3 - 1]} = \frac{0.1597}{0.1639} = 0.974$$

$$F^\circ_{corr} = F_m \cdot \frac{T_C}{T} \cdot \frac{P - P_{H_2O}}{P} \cdot j = 1.99\,mL/min \times \frac{423.15\,K}{295.15\,K} \times \frac{0.970\,atm - 0.026\,atm}{0.970\,atm}$$

$$\times\, 0.974 = 2.70\,mL/min$$

In this equation, the vapor pressure of water was converted to atmospheres to keep units consistent.

$$V^\circ_R = t_R \cdot F^\circ_{corr} = 5.43\,min \times \frac{2.70\,mL}{min} = 14.7\,mL$$

$$V^\circ_M = t_M \cdot F^\circ_{corr} = 0.813 \times \frac{2.70\,mL}{min} = 2.20\,mL$$

$$V_N = V^\circ_R - V^\circ_M = 14.7\,mL - 2.20\,mL = 12.5\,mL$$

Repeat the calculation of the net retention volume for a solute that elutes at $t = 7.58\,min$ under the following conditions: 35% diphenyl 65% dimethyl polysiloxane column; column length, 60 m; column i.d. (inner diameter), 0.180 mm; column temperature, 325 °C; ambient temperature, 21.0 °C; inlet pressure, 1189 mmHg; ambient pressure, 0.983 atm. An unretained species eluted at $t = 1.254\,min$ under the same conditions. The vapor pressure of water at 21.0 °C is 18.6 mmHg. The column flow rate was measured using a soap bubble meter and found to be 3.04 mL/min.

**Answer:**

$$j = 0.7585$$

$$F^\circ_{corr} = 4.62\,mL/min$$

$$V_N = 35.02 - 5.79 = 29.23\,mL$$

## 2.3.   GAS CHROMATOGRAPHY INSTRUMENTATION

Gas chromatography, as stated earlier, is based on the partitioning of molecules between a gaseous mobile phase and a liquid stationary phase. When considering the fundamental concepts that have already been presented in this chapter, it is clear that a gas chromatograph must be designed to accomplish the following:

1. Introduce the sample to be analyzed;
2. Provide carrier gas flow that transports samples through the injection port (where the sample is introduced into the instrument) and the column;
3. Control the temperature of the column, including the ability to systematically vary the temperature during the course of the analysis;
4. Detect the molecules as they elute from the end of the column; and
5. Convert the signal into data that can be analyzed.

A schematic of a gas chromatograph is shown in Figure 2.8. The functions of each component are discussed in the following sections.

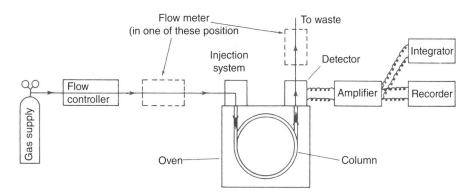

**FIGURE 2.8**　Schematic of a gas chromatography system. (*Source:* Gas Chromatography, J.E. Willett, John Wiley and Sons, Inc. 1987 page 10, Reproduced with permission.)

Before discussing the instrument, a brief word about the sample is required. The sample to be analyzed (e.g., perfume or oil) is commonly dissolved in a volatile organic solvent such as acetone, methanol, or an alkane. Thus, the solute molecules themselves often represent a very small fraction of the sample that is introduced into the instrument. In some cases, pure sample is injected if solute concentrations are expected to be low or if the sample itself is mostly solvent. The important factors, however, are that the samples must be at least somewhat volatile (i.e., possessing boiling points roughly below 350 °C) in order to elute from the column and they must be thermally stable at the temperatures employed in the instrument.

### 2.3.1.　Carrier Gas Supply

Carrier gas is typically provided by high-pressure tanks connected to the sample introduction chamber (injection port) via metal tubing. High-purity (99.995% or higher) He, $N_2$, and $H_2$ are commonly used carrier gases. Regulators on the tank help maintain appropriate working pressures and indicate the amount of gas left in the tank. The inlet to the instrument controls the total flow of gas supplied to the injection port and to the column. In modern instruments, these flows are set electronically. In older instruments, the flow is controlled using pressure regulators built into the instrument. Gauges are present on the instrument to indicate the pressure being applied to the head of the column (referred to as the head pressure).

As discussed earlier, the nature of the inert carrier gas and the flow rate affect the efficiency (plate height) of the column and the retention of the solutes. Therefore, it is important to know and record the flow rates that are used so that methods and analyses can be repeated in the future.

### 2.3.2.　The Injection Port and the Solute Injection Process

Problems arising in GC separations can often be traced back to the injection process. For this reason, the injection process is the most critical step to understand in order to obtain reproducible results and to maximize the performance of the system. Detailed books and articles have been written about the injection process and how it affects quantitative analyses.[7-11] It is recommended that these resources be read and understood by those wishing to use GC as an analytical technique. An overview of the injection process and different modes of introducing samples into the column follows.

A schematic of a GC injection port is shown in Figure 2.9. The injection port is usually held at a high temperature (250–300 °C) and is lined with a fused silica liner that is chemically derivatized to deactivate its surface. The liner provides open space for the sample to volatize before entering the column. It also prevents solutes from touching metal surfaces inside the injection port that could catalyze decomposition reactions. A variety of glass liners are available and each provides subtly different injection dynamics. Discussion of those details, however, falls outside the scope of this text, but good resources describing the different configurations are available.[9-13] In most GC injection systems, the column enters the liner at the very bottom (see Figure 2.9).

The sample is introduced into the heated injection port using a syringe. This is done manually or with an autoinjector. The syringe is pressed through a polymeric septum. For open tubular systems, typically 0.1–2 µL of sample are injected. Once the needle is inserted into the injection port, the plunger is rapidly depressed, expelling the liquid sample into the injection port. The sample is quickly vaporized due to the high temperature and heat capacity of the injection port. This method of injection is therefore referred to as *flash vaporization*. After the injection process, the syringe is quickly removed. It is critical that the

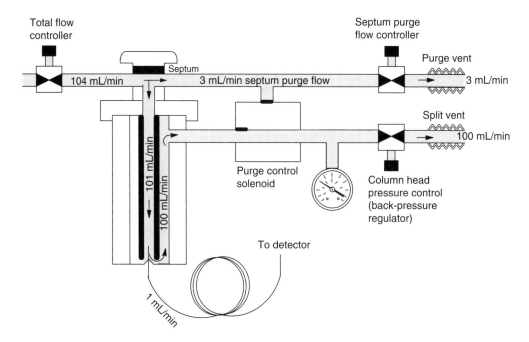

**FIGURE 2.9**  Schematic of split/splitless injection port. (*Source:* © Agilent Technologies, Inc., 1996, Reproduced with permission, Courtesy of Agilent Technologies, Inc.)

entire injection process be done quickly, for once the needle enters the hot injection port, the solvent and solutes begin to vaporize and enter the column. If the injection is done slowly, the sample is introduced over a relatively long period of time. This leads to severe band broadening and degrades the resolution of the separation.

Throughout the injection and the remainder of the separation, carrier gas is introduced through the injection port and into the column. The flow rate of carrier gas through the column is controlled by several valves. As solutes are vaporized in the injection port, they are swept onto the column where they separate based on the partition process described earlier.

The diagram of the injection port shows that the carrier gas, in addition to flowing onto the column, can also flow through the septum purge vent. This flow across the bottom of the septum reduces band broadening that could occur if some sample liquid on the outside of the syringe is wiped off on the underside of the septum as the syringe is retracted from the injection port. If the septum purge flow were not present, the solutes in that liquid would evaporate from the underside of the septum and *slowly* make their way onto the column, causing band broadening. With the septum purge flow, any molecules that desorb from the septum are swept through the purge vent and exit the instrument instead of being carried onto the column. A typical carrier gas flow rate through the purge vent is 3 mL/min.

It is important to note the significant expansion of liquids as they vaporize. Table 2.3 lists several common GC solvents and the volume of gas that 1 μL of the liquid expands to at 300 °C and a typical injection port pressure of 20 psig.[11] Given that injection ports typically have volumes on the order of 1–2 mL, it is clear that the sample can overload the capacity of the injection port if too much liquid is injected. When this occurs, solute molecules are chaotically swept into other parts of the injector and also exit through the septum purge vent. This is referred to as *flashback* and causes loss of solute, which can be detrimental to the analysis. In addition, the reproducibility of peak heights and areas upon which reliable quantitation depends suffers significantly. It is therefore important to be aware of injection volumes, the expansion of liquids, and the complex nature of the injection process in order to achieve reliable quantitative analyses.

There are many different ways of injecting solutes into a column. Most of them involve injecting the sample into the injection port liner rather than onto the column directly. They differ in the temperature of the injection port at the time of injection and in whether the split vent is opened or closed during injection. The main characteristics of the different injection modes are summarized in Table 2.4 and discussed in detail in the following sections.

**TABLE 2.3 Volume of Gas Occupied by 1 μL of Liquid Solvent**

| Solvent | Gas volume (mL) |
| --- | --- |
| Ethyl acetate | 0.203 |
| Hexane | 0.152 |
| Methanol | 0.503 |
| Methyl *t*-butylether | 0.167 |

**TABLE 2.4   Characteristics of Common GC Injection Modes**

| Injection mode | Injection port settings | Advantages | Disadvantages |
| --- | --- | --- | --- |
| Split | Hot. Sample injected into the liner. Split vent open to expel a specified ratio of the injected sample | Prevents overloading the column, which leads to poor peak shapes. Good for concentrated samples and thin stationary phase films | Some sample is "wasted" (i.e., not analyzed), so it may not be appropriate for trace analysis |
| Splitless | Hot. Sample injected into the liner. Split vent closed during injection and then opened after a specified time to sweep out residual vapors | Minimizes solvent tailing that can overlap early-eluting compounds. Almost all of the sample enters the columns. Good for trace analysis | Could lead to overloading and consequently to poor peak shapes for concentrated samples or thin stationary films |
| Direct | Similar to splitless, but the split vent is never opened | All of the sample enters the column making it a good technique for trace analysis | Poor peak shapes and large, tailed solvent peaks may result |
| Cool on-column | Initially cool and then heated after the sample is deposited directly into the column | Potentially better reproducibility and less solute discrimination than injections made into the injection port. Good for trace analysis since the entire sample is deposited into the column | Takes time to heat and cool the injection port so the time per analysis increases. Difficult to insert the syringe into a capillary column and potentially damaging to the column |
| Programmable temperature vaporization | Cool. Sample is injected onto a plug of glass wool in the liner. The injection port is then heated to vaporize the solutes and solvent. Split vent can be closed or open depending on desired effects | Can manipulate the split vent to preferentially eliminate the solvent but retain the solutes. Allows for larger sample volumes to be injected without interference from the solvent peak. Good for trace analysis | Requires time to cool and heat the injection port so the time per analysis increases |

**EXAMPLE 2.6**

When using heated injected ports, (a) how are peak widths affected if the syringe plunger is depressed slowly rather than rapidly and (b) why is it desirable to have the temperature of the injection port above the boiling points of the solutes and solvents?

**Answer:**

(a) If the syringe plunger is depressed slowly, solutes are introduced over a wide time range. This means that each peak elutes over a broader time (i.e., a broader peak), which leads to lower resolution. It is therefore desirable to introduce the solutes as quickly as possible when using heated injection ports.

(b) If the temperature of the injection port is lower than that of the solvent or solutes, they will condense in the injection port. If this occurs, they are not carried into the column because the main mechanism by which solutes enter the column in most modes of injection is by being swept into it by the carrier gas flowing through the injection port. This would lead to poor quantitation for those solutes that do not enter the column.

### 2.3.2.1. *Modes of Injection: Split Injection.* Injecting too much sample into capillary columns often leads to poor peak shapes and poor resolution. Therefore, injections are often conducted in what is referred to as split injection mode. In split injection mode, flow control valves within the instrument divide the total carrier gas flow between the column and the split vent. The ratio of the flow through the split vent relative to that through the column is called the split ratio. It dictates the amount of sample that enters the column. For example, if 40 mL/min of carrier gas are going through the split vent, and 2 mL/min through the column, the split ratio is 20:1. In this case, only 1 out of every 20 molecules that are injected actually makes it onto the column to be analyzed. The rest are swept out of the instrument through the split vent. In this way, sample sizes are reduced so as to not overload the column and thus obtain good resolution and peak shapes.

It is logical to ask: "Why not inject less sample?" The answer to this is that typically, only 1 μL is injected to begin with. It is quite difficult to *reproducibly* handle and inject fractions of a microliter. Thus, it is easier and better analytically to manipulate larger samples and use the instrument flow controls to achieve the desired injection amounts.

Typical split ratios vary from 2:1 to 100:1 depending on the analysis being conducted and the nature of the sample. The split ratios are achieved by adjusting both the valve that controls the flow through the split vent and the valve that controls total flow into the instrument.

For example, to achieve a flow rate through the column of 5 mL/min and a split ratio of 10:1, the total flow into the injection port needs to be 58 mL/min (5 mL/min through the column, 50 mL/min through the split vent, and 3 mL/min through the purge vent).

While split injections provide improved peak shapes and resolution, the majority of the sample goes undetected and is wasted. In a sense, by using split injection, the challenge of detecting the molecules in the sample has been greatly increased. For relatively concentrated samples, this is not an issue. For trace solutes, however, split injection may not be practical.

### 2.3.2.2.   Modes of Injection: Splitless Injection.

In splitless injection, flow through the split vent is temporarily stopped at the beginning of the injection process and all of the carrier gas flowing through the injection port goes onto the column, taking with it the majority of the sample. This is often necessary for trace analyses when wasting sample is not an option. In a typical splitless injection, the split vent is closed at the time of injection and remains closed for a designated period (typically 30 s to 2 min). After the designated time, the split vent is opened in order to purge any remaining sample in the injection port. This is done to minimize the tailing that arises from the large solvent peaks that can overlap early-eluting species as shown in Figure 2.10.

---

**EXAMPLE 2.7**

Which injection method, split or splitless, is more appropriate for the trace analysis of an impurity in a solvent that has a retention time similar to that of the solvent?

**Answer:**
Splitless injection would be more appropriate to try first because it is a *trace* analysis. Split injection offers the benefit that the solvent peak area would be greatly reduced and thus makes it possible to resolve it from the closely eluting impurity. However, if the split vent were open during the injection, as it is in split injection, the analyte may be nearly completely expelled and therefore undetectable. In splitless injection, more of the analyte may make it onto the column, but the solvent peak width and area may be diminished enough that the impurity can be resolved from it.

---

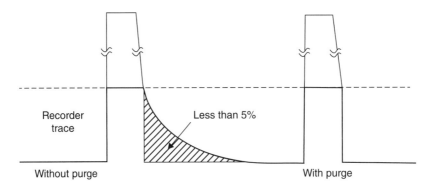

**FIGURE 2.10**   Elimination of solvent tailing using inlet purge. Without purge, there is the potential for the solvent peak "tail" to overlap solute peaks, making quantitation more difficult or impossible. (*Source:* © Agilent Technologies, Inc., 1991, Reproduced with permission. Courtesy of Agilent Technologies, Inc.)

***2.3.2.3.   Modes of Injection: Direct Injection.*** Direct injection is similar to splitless injection except the split vent is never opened. Therefore, all of the sample makes it onto the column after being vaporized in the hot injection port. This mode is more commonly used in instruments that are manufactured for packed bed columns and subsequently modified to accommodate megabore capillary columns (0.53 mm in diameter). Again, the advantage is that no sample is wasted, so peak areas are maximized, making detection of low concentration solutes possible. However, poor peak shapes and broad solvent peaks may result from direct injection of large sample volumes. Direct injection is often confused with on-column injection discussed below. In both cases, the entire sample makes it onto the column. However, in direct injection, the sample is injected and vaporized in the injection port liner *prior* to entering the column. In on-column injection, the syringe is inserted into the column itself where the injection is made.

***2.3.2.4.   Modes of Injection: Cool On-Column Injection.*** In cool on-column injection, the injector and the head of the column are held at a temperature below the boiling point of the solvent while the liquid sample is injected directly into the column using a syringe. The inlet and column are then heated to achieve vaporization of the sample components to begin the separation.

The advantage of this mode of injection is that the injections can be more reproducible than with flash vaporization techniques because the liquid sample itself is injected directly into the column without any vaporization or subsequent expansion. This also substantially reduces discrimination in the injection process. Discrimination issues are discussed below. Finally, because the entire sample is introduced onto the column, no sample is wasted as it is in split and splitless injection.

The disadvantages are that the heating and cooling of the inlet require time, which increases the analysis times, inserting syringe needles into columns smaller than 0.53 mm in diameter is difficult, and damage to the syringe and column can result if done improperly.

***2.3.2.5.   Modes of Injection: Programmable Temperature Vaporization Inlet.*** Programmable temperature vaporization (PTV) inlets combine the advantages of cool on-column injection and split/splitless injections. Liquid samples are injected into a cool injection port liner, typically containing a plug of glass wool or other surface onto which the sample is deposited. The temperature of the injection port is then increased and sample components enter the gas phase according to their boiling points. Depending on how the split vent is controlled during this time, multiple methods of injection are possible. If the solvent is low boiling, the split vent can be open at the low temperature when the solvent is coming off the glass wool. This ejects the majority of solvent through the split vent while the analytes remain in the injection port liner. At higher temperatures, the split vent can be closed so that the majority of the analyte molecules are delivered to the column upon vaporization from the liner. This method significantly reduces the size of the solvent peak and therefore makes possible the analysis of early-eluting compounds that might otherwise be obscured by a large solvent peak.

This method is particularly advantageous for very dilute samples because it allows for large sample volumes to be injected. Solvent interference is minimized while analyte signals are maximized.

**Cold Split PTV Injection.** In cold split PTV injection, the sample is deposited into the cool liner. The split vent is opened and the temperature of the inlet is increased. Thus, all components of the sample experience identical split ratios as they sequentially desorb from the liner according to their boiling points. This mode offers advantages over flash vaporization split injections by making it possible to inject more sample without significant increases in band broadening due to injection phenomena. Furthermore, the reproducibility of the analysis is improved due to a more controlled sample introduction into the column.

**Cold Splitless PTV Injection.** In cold splitless PTV injection, the sample is introduced in liquid form into a cool liner that is subsequently heated to vaporize the sample components in order of increasing boiling point. The split vent is opened near the end of the injection (30–90 s after beginning the injection) to expel residual vapors. The majority of the injected sample makes it onto the column, making this an attractive mode for the analysis of trace analytes.

### 2.3.2.6. Injection Discrimination.
As stated earlier, while GC is a simple technique and gas chromatographs are relatively simple instruments, problems can and do arise, and quite often these problems have their source in the injection process.

One common phenomenon observed is discrimination in which some solutes are preferentially transferred onto the column. This is common with flash vaporization using manual injection. In what is known as *filled needle injection*, sample is drawn into the tip of the syringe, which is then inserted through the septum into the hot injection port. The volatile components of the sample, including the solvent, immediately begin vaporizing, leaving the less volatile components behind in the syringe.

One method for reducing this discrimination is to draw the sample into the needle followed by drawing a plug of air into the syringe needle. This serves to draw the sample into the body of the syringe, leaving the needle filled with just air. The needle is then inserted into the hot injection port and a delay of 3–5 s is employed before depressing the syringe. This allows the syringe needle time to warm up and thereby reduces the discrimination. This is referred to as *hot needle injection*.

The injection method with essentially no discrimination is cool on-column injection. As already described, the entire sample is injected directly into the column as a liquid without a vaporization step. Figure 2.11 shows the discrimination of high-boiling (low-volatility) *n*-alkanes that are introduced using flash vaporization methods compared to cool on-column injections.

The discussion above pertains largely to manual injections because even average fast manual injections takes a few seconds to complete, which provides enough time for needle discrimination to occur. Autosamplers inject samples much more quickly and are therefore less prone to needle discrimination effects.

### 2.3.2.7. Inlet Discrimination.
The injection port itself can cause discrimination of both high and low boiling solutes. If too much sample is injected (keep in mind the dramatic liquid-to-gas expansion that occurs that is detailed in Table 2.3), the volume of sample vapors can expand beyond the volume of the inlet liner causing flashback, which pushes sample molecules out of the inlet and into surrounding areas such as the septum purge vent. In this case, the most volatile sample vapors can be pushed to the top of the

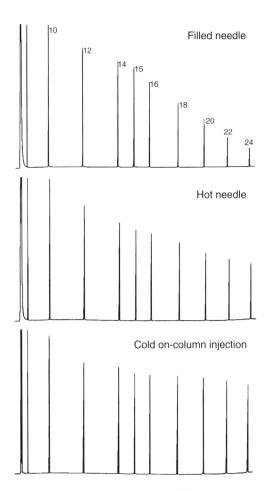

**FIGURE 2.11**   Discrimination of *n*-alkanes up to $C_{24}H_{50}$ for filled needle and hot needle injection compared to cold on-column injection. Note the discrimination that occurs for the higher boiling *n*-alkanes with filled needle injection. With cold on-column injection, all of the solutes are delivered to the column and thus no discrimination based on the relative volatilities of the solutes occurs. (*Source:* K. Grob, and R. Grob, Practical Capillary Gas Chromatography – a Systematic Approach; *J. High Res. Chromatogr. & Chromatogr. Comm.*, 1979 (2), 109–117. Copyright Wiley-VCH Verlag GmbH & Co. KGaA. Reproduced with permission.)

injection port, which is cooler than the other surfaces because it is in contact with ambient temperatures. There, they can condense in the cooler region of the injection port and on surfaces such as the septum or gas lines. Other sources of discrimination for highly volatile solutes include long residence times of vapors in the inlet, high inlet temperatures, and low inlet pressures.

High boiling analytes may adsorb onto the inner surfaces of the injection port. If this happens, increasing the inlet temperature will result in increased peak areas for late-eluting peaks. By systematically varying the temperature and pressure of the inlet, one can check for increases and decreases in signal of early- and late-eluting compounds. If the peak areas change as a function of the inlet pressure and temperature, then discrimination effects are likely present.

Solute discrimination has significant implications for quantifying solutes and can also be difficult to reproduce, leading to poor analytical reliability. It is therefore important that practicing chromatographers be aware of these potential complications and use appropriate controlled experiments to test for the presence of bias in their injection methods.

### 2.3.2.8. Solvent and Stationary Phase Focusing Effects.

Through judicious selection of solvents, stationary phases, and initial column temperatures, the injection process can be used to focus solutes, especially low boiling solutes, into narrow bands immediately after injection. In stationary phase focusing, the initial column temperature is set significantly below the boiling point of the most volatile solutes in the mixture being analyzed. As solutes exit the hot injector in the vapor phase, they immediately condense at the head of the much cooler stationary phase. This can be pictured like the stacking of cars that occurs in an accident zone. Cars that had been well spread out in time and space are concentrated into a narrow band. Similarly, solutes that were far apart in the injection port and entered the column at different times are focused into a more concentrated solute zone. Then, as the temperature of the column increases using a temperature program, the solutes travel down the column in a much narrower band than they would have if the initial zone dispersion were not reduced.

Solvent focusing occurs when the initial column temperature is set well below the boiling point of the solvent in which the sample is dissolved. As the solvent enters the column, it condenses and essentially floods the stationary phase. The analytes in the sample are condensed in this flooded zone. As the temperature of the column is increased, the volatile analytes are concentrated at the tail end of the solvent as it evaporates. Thus, high-volatility components are focused. This process is depicted in Figure 2.12.

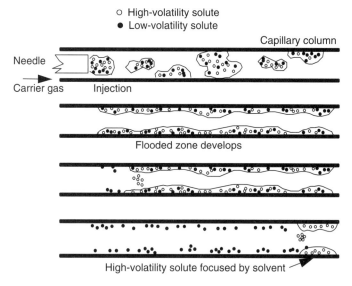

**FIGURE 2.12** Solvent focusing arising from the condensation of the solvent on a cool column followed by evaporation as the column temperature is increased. (*Source:* © Agilent Technologies, Inc., 1991, Reproduced with permission, Courtesy of Agilent Technologies, Inc.)

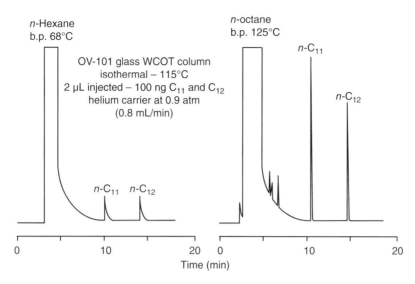

**FIGURE 2.13** Chromatograms showing the effect of solvent focusing. On the left, no solvent focusing occurs because the solvent boiling point is lower than that of the column temperature. On the right, however, the peaks for undecane ($n$-$C_{11}$) and dodecane ($n$-$C_{12}$) are sharpened due to solvent focusing induced by using a solvent, $n$-octane, which has a boiling point (125 °C) that is higher than the temperature of the column (115 °C). Note that on the right, not only are $n$-$C_{11}$ and $n$-$C_{12}$ focused, but minor peaks, possibly arising from solvent impurities, are also focused such that they can be observed on the right but disappear into the solvent tail on the left. (*Source:* © Agilent Technologies, Inc., 1991, Reproduced with permission, Courtesy of Agilent Technologies, Inc.)

Figure 2.13 shows a chromatogram in which no solvent focusing occurs due to the initial column temperature being higher than the boiling point of the solvent (and therefore no condensation, no flooded zone, and no focusing), compared to a chromatogram in which a higher boiling point solvent was used to achieve focusing. Note that all other conditions (column type, stationary phase, etc.) are the same. It is clear that solvent focusing can greatly improve chromatographic separations and yield more reliable analytical results.

***2.3.2.9. Retention Gaps.*** Figure 2.14 shows that using a length of capillary column that has been deactivated but that does not contain stationary phase, called a retention gap, provides a surface on which solvent focusing can be achieved. Then, as the solvent exits the retention gap and encounters the stationary phase, stationary phase focusing may also occur. Retention gaps are also useful for trapping nonvolatile solutes before they enter the stationary phase and permanently alter its chemical nature. Because the retention gaps are short (much less than a meter) and inexpensive, they can be replaced often in order to protect the much more expensive columns that are coated with the stationary phase, which typically cost several hundred dollars.

***2.3.2.10. Injection Valves and Sample Loops.*** The previous injection techniques involve discrete samples that have been collected, isolated, and then taken to the instrument for analysis. At times, however, it is necessary to sample from a flowing gas or liquid

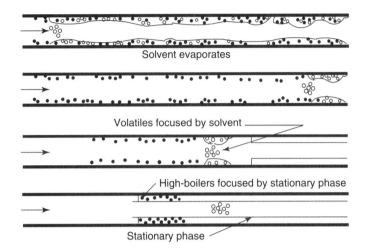

Solvent evaporates

Volatiles focused by solvent

High-boilers focused by stationary phase

Stationary phase

**FIGURE 2.14** Solvent focusing achieved using a retention gap coupled with stationary phase focusing effects. (*Source:* © Agilent Technologies, Inc., 1991, Reproduced with permission, Courtesy of Agilent Technologies, Inc.)

stream to get real-time analysis of a chemical process. This is especially useful in industry for continuous monitoring of chemicals. In these cases, sampling valves can be used in place of syringes to introduce gases and liquids into a gas chromatograph. A continuous flow stream of the sample to be analyzed is connected to a six-port sampling valve as shown in Figure 2.15. In Figure 2.15a, the sample flows into port #1, then through a channel in the valve to port #6, through a sample loop of fixed volume (typically 0.25–1 mL) to port #3, through another channel to port #2, and out to waste. All the while, carrier gas is being introduced at port #5, flowing through a channel to port #4, and entering the column. As depicted in Figure 2.15a, the carrier gas is completely bypassing any interaction with

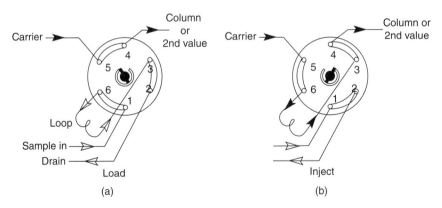

(a)                                                      (b)

**FIGURE 2.15** Six-port sampling valves. Configuration (a) is used for sample loading. Note that the sample loop is isolated from the carrier flow (i.e., mobile phase) so that sample can be loaded into the loop. The valve is then rotated to configuration (b) for sample injection. This puts the sample loop into the path of the flowing mobile phase and sweeps the sample onto the column. (*Source:* © Agilent Technologies, Inc., 1991, Reproduced with permission, Courtesy of Agilent Technologies, Inc.)

the sample. No analysis is being done in this configuration, yet the process is continuous because both the sample and carrier gas are flowing.

To actually analyze a sample, the valve is rotated, which causes the channels in the valve to be connected to different ports as depicted in Figure 2.15b. In this configuration, the carrier gas entering port #5 is now swept through a channel to port #6 where it encounters the sample loop, which contains the sample because sample had been flowing through the loop just prior to the valve switch. The sample is pushed out of the loop by the flowing carrier gas to port #3, which is now connected to port #4, which is connected to the GC column. In this way, the material in the sample loop is injected into the column and an analysis can be conducted. Meanwhile, the sample is still continuously flowing from port #1 to port #2. After the analysis is completed, the valve is returned to configuration A, which allows the sample loop to fill again in preparation for another analysis.

The fixed volume of the sample loop means that injection volumes are reproducible so that variations that occur in the peak areas are not due to injection irreproducibility but rather to chemical changes occurring in the process being monitored.

This discussion is appropriate for gaseous samples. For liquid samples, sample loops are not used because injecting 0.25–1 mL of liquid would overwhelm a typical GC column. To sample liquids, the valve channels themselves act as the sample loop, which limits the amount of liquid collected to volumes that are appropriate for GC analysis. The process of sampling and injecting, however, is the same as described for gas sampling.

***2.3.2.11. Purge and Trap.*** Many samples of interest are aqueous. Injecting water into GC columns, however, significantly reduces their lifetime because the water reacts with the silane linkages that bond the stationary phase to the column walls. This causes loss of the stationary phase, which decreases retention in the column. Aqueous samples are also problematic in that many organic species of interest are not highly soluble in water. Thus, even if it were advisable to inject aqueous samples into a gas chromatograph, solute peak areas may be too small to detect or quantify.

In many cases involving aqueous samples, a technique known as purge and trap sampling is used. In this technique, helium is bubbled (purged) through a large volume of the aqueous sample. Volatile organic components in the aqueous sample are swept up with the helium and thereby slowly removed from the sample. The helium flow is directed through a bed of adsorbent material (called the trap), where the organic components condense out of the gas phase onto the adsorbent. This concentrates the volatile organic components from the aqueous sample. The purge is stopped and carrier gas is directed through the trap. The trap is then rapidly heated to desorb the organic compounds. This flow of carrier gas and analytes is directed onto a GC column, where the trapped species are separated and quantified. Because the organic material is collected from a large volume of the aqueous sample (milliliters to liters), the sensitivity of purge and trap analyses is considerably better than if a mere microliter of the sample were injected directly onto the column. This technique is particularly useful in environmental analyses, where aqueous samples from rivers, lakes, and aquifers are common and trace organic compounds are the species of interest. Furthermore, it is important industrially because many chemical production processes use water as a solvent.

***2.3.2.12.  Headspace GC.*** A simpler, although less sensitive, alternative to purge and trap analysis is headspace gas chromatography (HSGC). The term headspace refers to the vapor phase above a condensed phase. In headspace GC, the vapor phase above a sample is sampled using either a manual gas-tight syringe or an automated headspace sampling unit. The vapor is injected onto a GC column, where the vapor phase components are separated and analyzed. The sample – typically an aqueous solution or solid – is sometimes heated to increase the vapor pressure of the volatile organic components present in it.

Headspace analysis is particularly well suited to odor analysis because our noses are reacting to molecules that are in the gas phase and therefore, by definition, volatile enough to be analyzed by HSGC. This is of great interest to the food industry because the smell of food often affects our overall experience of eating it, to the perfume and cologne industry, and to those who look for odor-removing adsorbents for footwear and baby diapers. HSGC is also employed in the analysis of residues in arson investigations.

***2.3.2.13.  Solid-Phase Microextraction.*** Headspace analysis can have low sensitivity because the concentration of analyte molecules in the gas phase is usually quite low, particularly if the analyte is highly soluble in the condensed phase below the headspace volume. Thus, it is sometimes desirable to include a step in the analytical process in which the analyte is concentrated prior to injection. This is referred to as preconcentration. In GC, one simple and convenient method of preconcentration is solid-phase microextraction (SPME – pronounced "speemee").[14,15] SPME achieves preconcentration by using a syringe-like device to introduce a solid silica fiber that is coated with an absorptive organic polymer film into the sample. Figure 2.16 shows an SPME device.

To use an SPME device, the silica fiber is withdrawn into the syringe needle using the spring-loaded plunger. The needle is then used to pierce a septum that seals the sample. Once the needle is inserted into the sample container, the plunger is used to push the silica fiber out of the needle in order to expose the adsorptive polymer film to the gas phase. Gas-phase analyte molecules are absorbed into the polymer film out of the gas phase. It is in this absorption step that the preconcentration occurs because relatively dilute gas phase molecules have been gathered into the small volume of the polymer film. When the system reaches equilibrium, the concentration of the analyte in the polymer film is proportional to the gas phase concentration, which in turn is proportional to the analyte concentration in the condensed phase sample.[14,15]

After equilibrium is reached, the fiber is again retracted into the needle. The SPME device is taken to a gas chromatograph and the needle introduced into the hot injection port. The fiber is then extended out of the needle. Due to the high temperature in the injection port, the analyte molecules desorb out of the polymer film and are swept into the GC column by the carrier gas flowing through the injection port. The key, however, is that more analyte molecules have been injected than would have been injected if the gas phase had been sampled directly without any preconcentration into the polymer film. Thus, the sensitivity of the analysis is improved.

A variety of polymer films are available, ranging from nonpolar polydimethylsiloxane (PDMS) to polar poly(acrylate) coatings. Different films have different extraction efficiencies depending on the analytes. In general, nonpolar coatings have higher efficiencies than polar coatings when used to extract nonpolar solutes and vice versa.

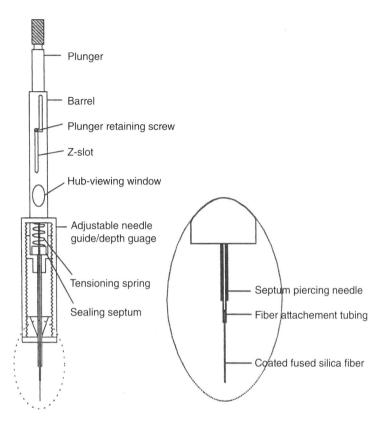

**FIGURE 2.16**   Solid-phase microextraction (SPME) device. The coated fused silica fiber is used to concentrate analytes from a gas or liquid sample. The fiber is then retracted into the needle. The needle is used to pierce the septum of a GC injection port. The fiber is then extended into the port. The solutes are thermally desorbed from the coated fiber in the injection port and swept onto the column by the carrier gas flow.  (*Source:* J. Pawliszyn, Solid Phase Microextraction: Theory and Practice, John Wiley and Sons, Inc., 1997, page 38. Reprinted with permission.)

While this discussion focused on gas phase analysis, SPME is equally capable of preconcentrating analytes in liquid samples. For example, many organic compounds have very low solubilities in water and are therefore hard to detect because of their low concentrations. Yet it is important to be able to detect even low concentrations of some organic species in water (e.g., for environmental purposes and for clean drinking water concerns). SPME can be used to preconcentrate organic species by inserting a nonpolar polymer-coated silica fiber directly into the aqueous phase being analyzed. The organic species partition out of the aqueous phase and into the organic polymer phase. The analyte molecules are then introduced into the GC in the manner described above.

### 2.3.3.   Oven/Column Compartment

Regardless of the injection process (split, splitless, headspace, SPME, etc.), once the sample has been injected and reaches the head of the column, the temperature of the column must be rigorously controlled in order to achieve the desired separation and to do so *reproducibly*. For this reason, the column is housed in a thermostated oven. The oven uses forced air to bathe the column at the desired temperature. Typical ovens can be reliably varied between 35 and 400 °C and controlled to within 1 °C. Cryogenic cooling units (using liquid $CO_2$ or $N_2$) can be added to the system to achieve temperatures down to −100 °C in the event that extremely volatile components need to be separated. There is little need for ovens to exceed 400 °C because most columns begin to decompose and degrade at temperatures above 400 °C. The user must be aware of the maximum allowable temperature of the columns as they vary between stationary phases.

Most GC ovens are also capable of temperature programming, with maximum heating rates of 20 °C/min. As mentioned earlier, the use of temperature programming can significantly increase the overall analysis time because the oven temperature must be returned from the final high temperature of the previous run to the initial low temperature of the next run. This is done by drawing ambient air into the oven while forcing the hot air out through open vents in the back of the oven. The greater the temperature difference and the closer the initial temperature is to ambient temperature, the longer it will take between analyses for the system to come to equilibrium at the initial temperature in preparation for the next analysis. However, some analyses take too long if conducted isothermally because the late-eluting solutes are not very volatile. In these cases, temperature programming is a powerful way to reduce the overall analysis time by using higher temperatures to drive off the well-retained solutes.

Because typical capillary columns are 10–60 m in length, in order to be housed inside the oven they are coiled around a wire frame referred to as a basket. The brown polyimide coating applied to the outside of the fused silica glass provides the mechanical stability and flexibility that allows for this coiling. Packed columns housed in metal or glass tubing are formed in coils to fit in conventional GC ovens. Unlike capillary columns, which are flexible and can be manipulated, packed columns are rigid and are typically not altered.

## 2.3.4. Detectors

In the previous sections, we saw how carrier gas flow is delivered to the injection port and the column. The next instrumental task is to detect the solutes as they elute from the end of the column. There are a variety of detectors available for gas chromatographs, each with their own strengths and limitations. They all take advantage of particular characteristics or properties of the analyte molecules: thermal conductivity, ability to burn in a flame, absorption of IR radiation, and absorption of electrons are all exploited to make detectors. Not surprisingly, then, the best detector for a given application depends on the analytes, their expected concentrations, and the information about the sample that is desired. As one eminent chromatographer summarized it: "The ideal chromatographic detector gives very good S/N for only the compounds of interest at the concentration you are working at, has an acceptably wide linear range, and is insensitive to changes in operating variables" (private communication with Professor Peter Carr, University of Minnesota).

In this regard, detectors are characterized according to their *detection limit*, *selectivity*, and *linear range*. The detection limit is the smallest amount of material that the detector can distinguish from noise. The selectivity reflects the ability of the detector to detect only specific analytes of interest. For example, some GC detectors detect nearly every compound, while others respond strongly to a specific class of compounds (e.g., polyhalogenated species) but are almost insensitive to all other compounds. Depending on the analyst's goals, this selectivity can be a positive or negative attribute of a detector. Finally, the linear range is a measure of the concentration or mass range over which the response of the detector is linear and therefore convenient for establishing calibration curves for quantitative analyses. Detectors with large linear ranges are favored because they allow samples with a wide range of concentrations to be analyzed using the same calibration curve.

The most common modes of detection for gas chromatography are flame ionization, thermal conductivity, electron capture, infrared, and mass spectrometry, although other methods also exist. The physics and operation of each are discussed below. Table 2.5 summarizes the detection limits, selectivity, and linear ranges of these detectors.

**2.3.4.1.** *Flame Ionization Detector (FID).* Because gas chromatography is amenable only to species that can be transferred to the gas phase, and many of these compounds tend to be organic in nature, a detector that is sensitive to organic compounds is desirable for many applications. Flame ionization is one of the most popular methods of detecting organic compounds. In a flame ionization detector (FID) (see Figure 2.17), solutes are swept through a detector component known as an FID jet as they elute from the end of the column. At the tip of the jet, the solutes pass through a flame created by the combustion of a hydrogen/air mixture. Thus, this detector requires that both $H_2$ and compressed air be supplied via external high-pressure tanks. As organic solutes burn in the flame, they create ions. These ions are collected at electrodes called collector plates, creating a current in the detector circuitry. The more solute that is present, the more ions are created, which in turn

**TABLE 2.5  Characteristics of Common GC Detectors[7,16]**

| | Type | Detection limit | Selectivity | Linear range |
|---|---|---|---|---|
| Flame ionization detector (FID) | Mass-sensitive | 1 pg C/s | Nonselective: responds to nearly all organic compounds | $10^7$ |
| Thermal conductivity detector (TCD) | Concentration-sensitive | 1 ng/mL | Nonselective: responds if thermal conductivity differs from carrier gas | $10^5$ |
| Infrared (IR) | Concentration-sensitive | 1000 pg | Compounds with molecular vibrations | $10^3$ |
| Mass spectrometry (MS) | Concentration-sensitive | 10 pg–10 ng | Tunable for any species | $10^5$ |
| Electron capture detector (ECD) | Concentration-sensitive | 10 fg/s (lindane) | Halogenated compounds | $10^4$ |
| Nitrogen phosphorus detector (NPD) | Mass-sensitive | 1 pg N/s  0.5 pg P/s | N,P-containing compounds | $10^4$ |
| Flame photometric detector (FPD) | Mass-sensitive | 50 pg S/s  2 pg P/s | P,S-containing compounds | $10^3$  $10^4$ |
| Photoionization detector (PID) | Concentration-sensitive | 5 pg C/s | Aromatics | $10^7$ |

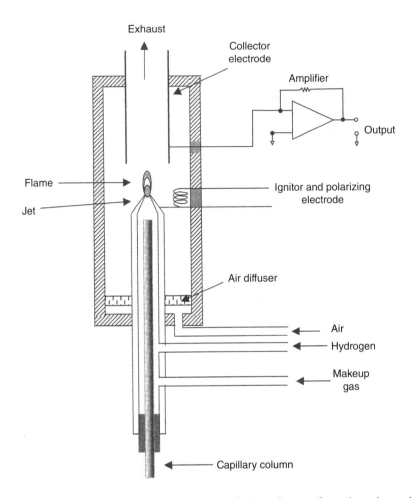

**FIGURE 2.17** Diagram of a flame ionization detector. Solutes flow up from the column through the detector. They burn in the flame and create ions. The ions are collected by the collector electrode and produce a current. A chromatogram is thus a plot of the detector current (or voltage) versus time. (*Source:* R.L. Grob and E.F. Barry, *Modern Practice of Gas Chromatography*, 4th edition, John Wiley and Sons, Inc., 2004, page 299. Reprinted with permission.)

produces more current. This current is then processed to create the chromatogram, which is essentially a plot of current versus time. When no solutes are eluting from the column, only the carrier gas is passing through the flame. Because the common carrier gases do not burn easily, the current in the absence of solutes is low and increases only when solutes pass through the flame. The sensitivity of FIDs depends on the ratio of $H_2$ and air flow rates. The user must systematically evaluate the system to achieve optimal sensitivity.

For organic species, the ions that form are mainly created by the breaking of C—H bonds. Thus, FID is a good detection method for solutes containing multiple C—H bonds. However, it is not a good detector for species that do not burn readily, such as $N_2$, $CO_2$, $CS_2$, highly halogenated molecules, and the rare gases. It is also not responsive to water, which is an advantage, given the difficulty of obtaining and maintaining absolutely dry gases and solvents.

When using an FID, it must be kept in mind that because of the way the detector works, its sensitivity is different for different molecules. For example, if the same number of moles of butane and decane are injected, the signal for decane will be larger because decane has more combustible C—H bonds per molecule. Furthermore, the relative sensitivity of FIDs to C—H bonds depends on the functional groups present in the molecule.[17,18] Because of this, if one observes the same peak areas for two species in a chromatogram, it *cannot* be assumed that the two components are present at the same concentration. In fact, it can almost be safely assumed that this is the least likely scenario. This is especially important to keep in mind when using GC to determine product ratios in organic synthesis. Each product must be independently calibrated before definitive statements about the relative percentages of compounds in the original sample can be made, particularly if the structures of the products are significantly different.

---

**EXAMPLE 2.8**

If equal concentrations of 2-butanone and 2-hexanone are detected using an FID, which peak will be larger and by what factor, assuming that signal is proportional to the total number of C—H bonds.

**Answer:**
The 2-hexanone peak will be approximately 1.5 times larger than that of 2-butanone. The factor of 1.5 arises from the relative number of C—H bonds (12/8). This illustrates that peak areas for different solutes cannot be directly compared to determine relative concentrations. It also illustrates the importance of using separate calibration curves for each solute.

---

It is also important to note that the FID is a *mass-sensitive detector*. This means that the *total area* under a chromatographic peak for a particular solute does not vary with the flow rate. Almost immediately upon entering the detector, the molecules are burned and ionized and the current created by the individual ions is recorded. The *rate* at which the molecules enter the detector affects the height and width of the chromatographic peaks, with higher flows leading to higher and narrower peaks, but the total number of ions created as the peak elutes, and hence the total integrated signal, does not depend on flow rate but only on the number of molecules (i.e., the *mass* of ionizable material) that enters the detector. It can be viewed simply that as each ion gets created, it gets "counted" once, so the rate at which the solutes pass through the detector does not affect the *total area* (but it will affect the height and width of the peak). This is in contrast to *concentration-sensitive detectors* such as the thermal conductivity detector (TCD) discussed in what follows, for which the total peak area depends on the through the detector. Sensitivities for mass-sensitive detectors are typically measured in nanograms per second, whereas sensitivities for concentration-sensitive detectors are often measured in nanograms per milliliter of mobile phase.

The issue of mass- versus concentration-sensitive detection impacts the use of GC for quantitative purposes. When using a mass-sensitive detector, a calibration curve based upon peak areas collected at one particular flow rate can be used to quantify a sample analyzed at a different flow rate. This is not true for concentration-sensitive detectors. It is also not true for calibration curves based upon peak height because peak height changes with flow rate in mass-sensitive detectors.

In operation, the temperature of the FID compartment is typically maintained at or slightly above the temperature of the injection port (if flash vaporization is used). Keeping the temperature in the FID compartment relatively high (approximately 250–300 °C is typical) reduces the potential condensation of high boiling solutes as they elute from the column. Condensation can foul the detector and decrease sensitivity.

When using FIDs, it is also common to supply an additional flow of helium at flow rates significantly greater than the column flow (20–40 mL/min) at the junction between the column and the FID jet. This so-called make-up flow facilitates the rapid transfer of solutes from the column to the jet. This reduces band broadening, which helps maintain resolution, and also increases peak height, which improves the signal-to-noise ratio.

### 2.3.4.2. *Thermal Conductivity Detector (TCD).* While the FID is commonly used to detect organic species, it suffers from its lack of sensitivity to polyhalogenated compounds and gases such as $N_2$, $O_2$, and $H_2$. Another widely used detector that can detect these as well as all organic compounds is the thermal conductivity detector (TCD).

A TCD contains a wire filament that has a constant voltage applied across it. The filament heats up and the heat generates resistance to current flow through the filament (similar to a light bulb filament). Effluent from the column is passed across and bathes the filament. Under constant flow of *pure* carrier gas (typically He), a constant temperature and hence constant resistance of the filament is established. As a solute band elutes from the column, the temperature of the filament changes because the solute almost always has lower thermal conductivity than helium. Consequently, solutes do not carry heat away from the filament as efficiently as the helium that was surrounding the element prior to the emergence of the solute. Thus, the temperature of the filament increases when a solute is present in the TCD, causing the resistance of the filament to increase. Because a constant voltage is being applied across the wire, as the resistance increases, the current through the wire must decrease (recall Ohm's law, $V = IR$). As a result, when a solute band passes through the detector, its presence is detected by a drop in the current through the filament. The chromatogram is therefore simply a plot of current versus time. The more solute that is present, the bigger the current change. Helium is commonly used in these detectors because it has a high thermal conductivity. Therefore, the difference between its thermal conductivity and the thermal conductivity of solutes is generally large, improving the sensitivity of the detector.

There are two common instrumental circuit designs for the TCD. One design is shown in Figure 2.18. In this design, a switching valve is used to alternately expose the filament to a reference flow of helium or to the column effluent. The valve is switched between the two flows every 100 ms. In one position, the reference helium flow is used to bathe the filament in helium and a baseline current is measured. When the valve switches, the column effluent then bathes the filament. If a solute is *not* present, then the helium from the column bathes the filament and there is no change in the filament temperature because

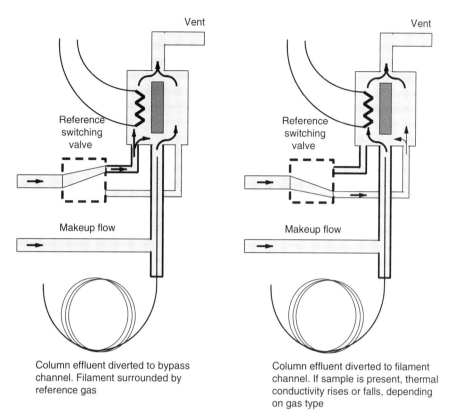

**FIGURE 2.18** Schematic of a thermal conductivity detector. When the valve is in the "up" position, carrier gas bathes the filament and column effluent is diverted away from the filament. In the "down" position, carrier gas sweeps the column effluent across the filament. If solutes are present, the thermal conductivity of the gas phase changes, leading to a change in the current in the filament. (*Source:* © Agilent Technologies, Inc., 1996, Reproduced with permission, Courtesy of Agilent Technologies, Inc.)

the reference and column flows have the same thermal conductivity. However, when a solute elutes from the column and passes through the filament zone, a new, lower current is measured. The difference between this current and that observed when only helium passes over the filament is plotted to create the chromatogram. This difference is measured with each switch of the valves, which, as stated above, occurs multiple times per second. In an instrument of this type, if you listen carefully, you can hear the switching valve toggling back and forth.

The sensitivity of TCDs is given by

$$S = kI^2 R \left( \frac{\lambda_c - \lambda_s}{\lambda_c} \right) (T_f - T_b) \tag{2.19}$$

where $k$ is a proportionality constant, $I$ is the current through the filament, $R$ is the resistance of the filament, $\lambda$ is the thermal conductivity of the carrier gas (c) and solute (s), and $T$ is the temperature of the filament (f) and the detector heating block (b). This equation shows that the detector is most sensitive to solutes with thermal conductivities that are significantly different than that of the carrier gas (i.e., greater $\lambda_c - \lambda_s$). To maximize this difference, it is best to use helium ($\lambda = 34.8 \times 10^{-5}$ cal/(cm s °C)) or hydrogen ($\lambda = 41.6 \times 10^{-5}$ cal/(cm s °C)) rather than nitrogen ($\lambda = 5.8 \times 10^{-5}$ cal/(cm s °C)) or argon ($\lambda = 4.0 \times 10^{-5}$ cal/(cm s °C)) because the thermal conductivities of organic compounds are typically below $15 \times 10^{-5}$ cal/(cm s °C).[19] This equation also shows that the response of the detector is greatest when the current through the filament is high and the temperature of the filament is much higher than the temperature of the heating block. There are practical limits to these parameters, however, that if exceeded significantly decrease the lifetime of the filament.

Another common TCD detection scheme is shown in Figure 2.19. Here, a Wheatstone bridge circuit is used to detect differences in the thermal conductivity of the column effluent and a reference helium flow. In a Wheatstone bridge, if all four resistance values are *constant*, a *constant* voltage difference between points $A$ and $B$ develops. However, if one of the resistances changes, the voltage between $A$ and $B$ changes. In this style of detector, resistors 3 and 4 ($R_3$ and $R_4$) are fixed resistors. Resistor 2 is bathed in a constant flow of the reference gas and thus it too would be constant. Resistor 1 is bathed in the column effluent. As solutes elute from the column, the resistance of Resistor 1 increases, resulting in a change in the voltage between $A$ and $B$. The chromatogram is therefore a plot of the voltage between points $A$ and $B$ versus time.

The advantage of the TCD is that it is a *universal detector*, meaning that it can detect nearly any solute that is not identical to the reference gas. The reason for this is that the principle of operation rests on the fact that the solute merely has to have a different thermal conductivity than the reference gas, and because, like fingerprints, no two species are likely to have absolutely identical thermal conductivities, a difference will always exist, making detection possible.

Another advantage is that TCDs do not require additional high-pressure tanks of $H_2$ and air that are necessary for FIDs. The same tank that supplies helium flow through the column also supplies the detector reference flow. In this way, TCDs are slightly easier and safer to operate than FIDs. However, the filaments can get fouled by reactions with organic

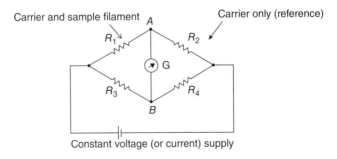

**FIGURE 2.19** Thermal conductivity detector based on a Wheatstone bridge. As solutes emerge from the column and are swept across the sample filament, the resistance changes due to the mismatch of thermal conductivities between solutes and the carrier gas. When the resistance changes, the voltage between $A$ and $B$ changes and in this way a chromatogram can be recorded by measuring the voltage as a function of time. (*Source:* © Agilent Technologies, Inc., 1991, Reproduced with permission, Courtesy of Agilent Technologies, Inc.)

species, and they can also burn out like light bulb filaments, meaning that they can require a bit more maintenance than FIDs. Another disadvantage of TCDs is that they are generally less sensitive to organic species than are FIDs.

Because TCD responses are dependent on the type of carrier gas used and the current through the filament, with optimum values depending upon temperature, it is important to consult the user's manual for optimization procedures when using a TCD. It is also important to note that TCDs are *concentration-sensitive detectors*, meaning that the *peak area* depends on the volume of mobile phase, and therefore the flow rate, that accompanies the solute molecules as they pass through the detector. As the flow rate decreases in a TCD, the solute molecules remain in contact with the filament longer and are therefore "detected" for longer periods of time. Thus, the signal integrated over time increases as the flow rate decreases. This is in contrast to mass-sensitive detectors like FIDs in which each solute molecule is destroyed by the detection process and hence cannot cause an accumulation of signal over time.

### 2.3.4.3. *Infrared Detector (IRD).* Fourier transform infrared spectroscopy serves as a valuable detection method for gas chromatography. It is especially valuable because IR is a nondestructive detection method, meaning that the molecules survive detection and can be collected or analyzed by a subsequent method. This is in contrast to techniques such as flame ionization in which the analyte molecules are ionized, chemically modified, or in some way obliterated, such that they do not survive the detection method.

In infrared detection (IRD), the column effluent is passed through a long, narrow tube called the light pipe. The walls of the light pipe reflect the IR radiation such that it has multiple opportunities to be absorbed by analytes. This is required because IR absorption is generally weak and multiple passes increases the probability that any individual photon is absorbed by the sample. By taking advantage of Fourier transform technology, the entire IR spectrum of an eluting compound is recorded multiple times per second. A chromatogram of absorbance versus time is thus produced. In addition, the IR spectrum collected at each moment of the analysis provides structural details about the individual components in a complex mixture. Furthermore, because the molecules escape the detector without structural modification, they can be passed on to a second detector, such as a mass spectrometer. The combination of an infrared detector and mass spectrometer can be used to *identify* specific compounds in the sample matrix, making this a good combination for qualitative analysis of mixtures.

An additional benefit of IR detectors is that they can function as both a *specific* and *universal* detector. For example, all organic molecules absorb IR radiation because they have bonds that can be excited by IR radiation. If, for example, information about ketones in a sample is desired, free from interferences from other components in the mixture, software can be used to show the chromatogram that results when only absorbance at $\sim 1715 \, \text{cm}^{-1}$ is recorded. Because only ketones absorb in this region, the chromatogram will reflect only substances with IR stretches at this wavenumber, as shown in Figure 2.20. A disadvantage of IR detectors is that some functional groups have small molar absorptivities in the IR region, leading to very weak signals for compounds lacking moieties that strongly absorb IR radiation.

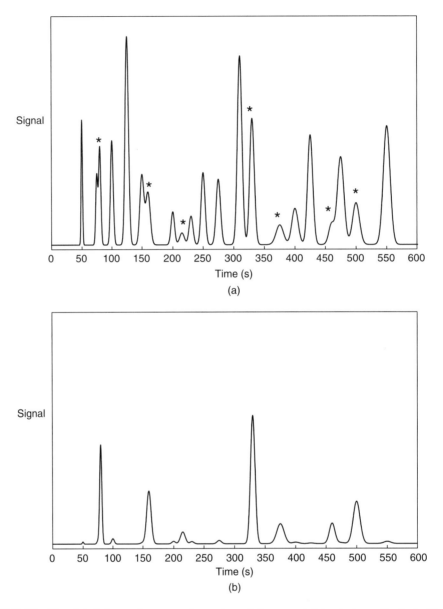

**FIGURE 2.20** Simulated chromatograms of complex mixtures containing ketones (indicated by asterisks). (a) Chromatogram resulting from integrating *all* IR signals (i.e., nonselective mode). (b) Chromatogram that might result by plotting only the signal observed at $1715\,cm^{-1}$ as a function of time.

### *2.3.4.4. Gas Chromatography–Mass Spectrometry (GC/MS).* Gas chromatography/mass spectrometry (GC/MS) is a popular and valuable analytical technique. In the most common mode of GC/MS, analytes are bombarded by a stream of high-energy electrons emitted from a tungsten filament as they exit the column in the gas phase. These electrons ionize the molecules in the effluent and cause bond rupture with subsequent molecular fragmentation. The charged fragments are subsequently passed through a

mass-to-charge analyzer (typically a quadrupole, magnetic sector, time-of-flight (TOF), or ion trap mass analyzer) and detected as a function of mass and time. Thus, a typical chromatogram for GC/MS is a plot of the total ion current detected by the mass analyzer versus time. However, it is also possible to select a specific mass-to-charge ratio ($m/z$) and monitor it alone. This is known as selected ion monitoring (SIM). For example, if one is concerned about benzene in a sample, the MS detector can be tuned solely to mass 78. In this mode, any solute that elutes that does not produce an ion of $m/z$ equal to 78 does not get registered in the chromatogram. Operating in SIM mode also improves sensitivity up to three orders of magnitude over the scanning mode because the instrument is using the time that would otherwise be used to sample undesired masses to sample the mass of interest.

Many instruments also allow for tandem MS studies, meaning that the ions that are separated in a first mass analyzer are then further fragmented through collisions with a gas. These collisions cause the initial ions to fragment further. These "daughter" fragments are then analyzed by another mass analyzer. The overall technique is known as GC–MS/MS, or more generally as GC–MS[n], where the superscript "$n$" denotes the number of fragmentation/detection stages that are used.

Mass spectrometers, like IR detectors, are universal because essentially *all* molecules create some MS signal, but they can be *specific* when tuned to detect only those molecules that produce ions of specific mass-to-charge ratios.

A schematic of a GC–MS interface is shown in Figure 2.21.[20] The instrumental challenge is to couple the column effluent that is coming out at several milliliters per minute to the mass selector that requires high vacuum in order for the ions to avoid collisions that would annihilate them. Furthermore, the ions must be focused and directed into the mass selector after being created by the electron ionization process. To accomplish these tasks,

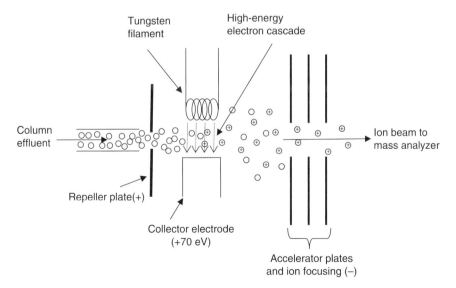

**FIGURE 2.21** Electron ionization GC–MS interface. Solutes are ionized by high-energy electrons emitted from the tungsten filament. The resulting ions are accelerated and focused by the accelerating plates and subsequently introduced into the mass analyzer. (*Source:* Adapted from Ref. 20.)

the interface is connected to a vacuum pump (not shown in Figure 2.21), and electrically charged plates direct and focus the ions. In addition to electron ionization (EI), it is also possible to use the softer ionization technique of chemical ionization (CI) in tandem with gas chromatographic separations.

It should be noted that while the above discussion is relatively short, there are a number of entire books on the principles of mass spectrometry and the wide range of ionization methods and mass analyzers that exist. The brevity of this section should not, therefore, be taken as an indication that GC/MS is not important. To the contrary, *GC/MS is routinely used and is one of the most powerful analytical techniques*. In the following chapter, we will see that mass spectrometers are also used as detectors in liquid chromatography (LC/MS), creating another powerful means for separating, identifying, and quantifying components in complex samples.

### 2.3.4.5. *Electron Capture Detector (ECD).*

In an electron capture detector (ECD), column effluent is passed through a stream of electrons produced by a radioactive source (usually $^{63}$Ni). Solutes passing through the detector absorb these electrons. In the absence of solutes, a constant current between the radiation source and a collector anode is developed. When solutes that absorb electrons are present, the measured current decreases. Thus, a chromatogram is created by measuring the current versus time.

Not all molecules capture electrons efficiently – both the number and type of atoms present are important. Compounds containing electronegative elements such as F, Cl, and Br are much better electron absorbers than those containing only C, H, N, and O. Table 2.6 shows the relative response of different compounds in ECDs. Clearly, this detector is best suited for the analysis of polyhalogenated organic compounds. As such, it finds wide use in environmental analysis, where many pollutants of interest are polyhalogenated species.

To efficiently transfer electrons to solutes, it is necessary to supply argon or nitrogen to the detector. These gases have large ionization cross sections and interact with the $\beta$ particles from the $^{63}$Ni source to create the secondary electrons that can then be absorbed by the solutes.

In contrast to flame ionization and thermal conductivity detectors, which can sense virtually all organic compounds, electron capture detectors are essentially insensitive to most

**TABLE 2.6  Relative Response of ECDs to Solutes[16]**

| Molecule | Response |
| --- | --- |
| Hydrocarbons | 1 |
| Ethers and esters | 10 |
| Aliphatic alcohols, ketones, amines, mono-Cl, and mono-F compounds | 100 |
| Mono-Br, di-Cl, and di-F compounds | 1,000 |
| Anhydrides and tri-Cl compounds | 10,000 |
| Mono-I, di-Br, poly-Cl, and poly-F compounds | 100,000 |
| Di-I, tri-Br, poly-Cl, and poly-F compounds | 1,000,000 |

nonhalogenated compounds. This may seem like a disadvantage: why would an analyst not want to detect all of the molecules that are present? It can, however, be a distinct advantage, especially for the analysis of complex mixtures in which only halogenated species are of interest. In this case, a detector that *selectively* detects only the molecules of interest greatly simplifies the analysis and reduces potential interference from overlapping signals from species that are not of interest.

### 2.3.4.6. *Nitrogen Phosphorus Detector (NPD).*

A nitrogen phosphorus detector (NPD, also called a thermionic detector, or TID) uses a heated rubidium silicate bead to selectively detect molecules containing nitrogen and/or phosphorus. Physically, the detector looks similar to an FID (see Figure 2.17). Air and hydrogen are used in this detector, as they are in an FID, but at considerably lower flow rates. In fact, no self-sustaining flame is created in this detector. Instead, in the heated detector, $H_2$ dissociates into reactive H atoms, creating an environment conducive to high-energy chemical reactions. The mechanisms of these reactions are not completely understood but rely on complex surface phenomena and the creation of a highly reactive gaseous layer near the hot rubidium bead.

As solutes enter this reaction layer, they are ionized and the ions are collected to create a current, producing a chromatogram of current versus time. Compounds containing N and P are selectively ionized, making the detector especially good for fertilizer, pharmaceutical, and chemical nerve agent analysis. With the proper control of operating parameters, these detectors can also be made selective for nitrogen or phosphorus independent of the other (i.e., operated in N mode or P mode).[21]

The NPD detector is similar to the ECD in that it is sensitive to a very narrow class of compounds. As is the case with the ECD, this is an advantage in that it can simplify potentially complex chromatograms, but does so at the expense of losing information about all of the other compounds in the sample mixture.

### 2.3.4.7. *Flame Photometric Detector (FPD).*

The flame photometric detector (FPD; see Figure 2.22) is in part similar to the FID, but instead of generating signal based on the collection of ionized species, detection is based on the emission of electromagnetic radiation from chemiluminescent species created in the detector. This detector is especially sensitive to compounds containing sulfur and phosphorus, as these elements readily form the chemiluminescent compounds required for detection. Sulfur- and phosphorus-containing species burn in the flame and generate the chemiluminescent species $S{=}S$ and HPO, respectively. These metastable molecules decay and emit light of specific wavelengths. The light is collected and passed to a photomultiplier tube, creating a current that provides the analytical signal. A 526-nm bandpass filter is used to detect P-containing species, and a 394-nm bandpass filter is used to detect S-containing species. This detector offers the same benefits and drawbacks as the other element-specific detectors discussed above in terms of its selectivity.

### 2.3.4.8. *Photoionization Detectors (PIDs).*

In a photoionization detector (PID), molecules are bombarded with UV electromagnetic radiation from argon or hydrogen emission lamps. The analyte molecules absorb the UV energy and subsequently eject an electron in order to dissipate the absorbed energy. The ionized molecules are collected at charged electrodes, creating a current that produces a chromatogram ultimately based

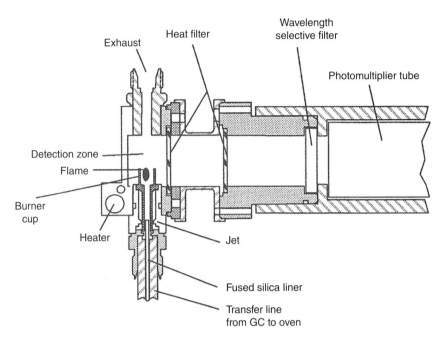

**FIGURE 2.22** Schematic of a flame photometric detector. Sulfur- and phosphorus-containing molecules burn in the flame to produce the chemiluminescent species S═S and HPO. The chemiluminescence is detected by the photomultiplier tube, the electrical output of which is used to produce the chromatogram. (*Source:* © Agilent Technologies, Inc., 1991, Reproduced with permission, Courtesy of Agilent Technologies, Inc.)

on current versus time. These detectors are particularly sensitive to molecules that ionize easily from UV light, such as aromatic and polyaromatic species and alkenes. Also, because only a small fraction of analyte molecules actually get ionized, PIDs are essentially nondestructive and can therefore be coupled with other detectors.

**Detector Summary.** There are many considerations when selecting a detector and the proper choice depends on the demands of specific analyses. Table 2.5, presented at the start of this section, summarizes some of the key aspects of the most common GC detectors.[7,16]

---

**EXAMPLE 2.9**

Which detectors discussed above are highly sensitive to specific elements, and to which specific elements are they sensitive?

**Answer:**
Flame photometric detectors (S + P)
Nitrogen phosphorus detectors (N + P)
Electron capture detectors (halogens)

## 2.4.  A MORE DETAILED LOOK AT STATIONARY PHASE CHEMISTRY: KOVATS INDICES AND MCREYNOLDS CONSTANTS

In the preceding sections, we first looked at how the nature and flow rate of carrier gases affect chromatographic analyses. We then explored different injection modes for introducing sample mixtures. Following this, we discussed the oven that controls the column temperature because temperature plays a significant role in governing solute retention and separation. We then looked at the advantages and disadvantages of a variety of detectors that respond to analytes as they elute from the column. All of these instrument components would be useless, however, if GC columns were unable to separate the individual components in complex mixtures. In the following section, we address the specific chemical interactions that occur between solutes and GC stationary phases and how they impact solute retention and separation. A consideration of these interactions guides column selection for the practicing chromatographer.

### 2.4.1.  Kovats Retention Indices

Measures of retention suffer from characteristics that make comparisons between laboratories difficult. For example, imagine that a laboratory reports the raw retention time for a specific solute, say benzene, on a specific column (e.g., a 0.53 mm i.d., 0.25 μm PDMS column). To reproduce the retention time, another laboratory would have to have *exactly* the same length column, with *exactly* the same film thickness, with *exactly* the same siloxane polymer, operating at *exactly* the same temperature, at *exactly* the same flow rate. In GC, this is simply not practical. Therefore, measures of retention that are independent of these variables are necessary to facilitate comparisons of solute retention obtained using different chromatographic systems and different operating conditions. Note that column length, film thickness, and flow rate do not alter the distribution constants of solutes – only temperature does (assuming the same polymer phase is being used). Therefore, if measurements of retention can be linked to the distribution constant, it should be possible to define retention measures that are independent of column length, film thickness, and flow rates. However, for comparisons to be made between two systems based on these retention measures, the temperature must be the same in both systems.

The retention factor, $k$, has already been discussed in Chapter 1. Given the definition

$$k = \frac{t_r - t_m}{t_m} = \frac{K}{\beta} \qquad (2.20)$$

it is clear that the retention factor depends on stationary phase thickness because of its dependence on the phase ratio ($\beta$) of the column. So in order for retention on different columns containing the same stationary phase to be compared, the phase ratio must be the same. The retention factor is, however, independent of column length and mobile phase flow rate. This can be understood if one considers retention of a solute on both 15 and 30 m columns. It will take the solute twice as long to elute from a 30 m column as it does from a 15 m column. But, it will also take an unretained solute twice as long to elute because it has twice the distance to travel. So both $t_r$ and $t_m$ are multiplied by 2, and the factor of 2 cancels as shown below:

$$15\,\text{m column: } k_{15} = \frac{t_{r,15} - t_{m,15}}{t_{m,15}}$$

$$30\,\text{m column : } k_{30} = \frac{2t_{r,15} - 2t_{m,15}}{2t_{m,15}} = \frac{t_{r,15} - t_{m,15}}{t_{m,15}} = k_{15} \qquad (2.21)$$

Similar considerations pertain to the effect of flow rate. Thus, $k$ as a measure of solute retention is independent of both column length and mobile phase velocity but is dependent on stationary phase thickness.

A measure of retention that is independent of the stationary phase thickness, as well as the column length and mobile phase flow rate, is the Kovats retention index. In order to understand the mathematical definition of the Kovats retention index, it is important to understand the retention behavior of the *n*-alkanes. The *n*-alkanes differ from one another simply by the number of methylene (—CH$_2$—) groups they contain. With each additional methylene group, retention of the *n*-alkanes increases due to the increased dispersion interactions that the additional methylene groups introduce. The important principle, first recognized by Martin,[22,23] is that each additional methylene group makes the same contribution to the total free energy ($\Delta G°$) of partitioning upon which retention is based. In other words, to a very good approximation, for an *n*-alkane with n carbon atoms (e.g., $n = 5$ for pentane),

$$\Delta G_n° = (n - 2)\Delta G_{CH_2}° + 2\Delta G_{CH_3}° \qquad (2.22)$$

where $\Delta G_n°$ is the total free energy of partitioning of an *n*-alkane with n carbons, $\Delta G_{CH_2}°$ is the free energy contribution per methylene group, and $\Delta G_{CH_3}°$ is a constant contribution from the terminal methyl groups.

The important aspect of this relationship is that *the free energy of partitioning is linearly related to the number of carbon atoms in the n-alkanes* because each methylene group interacts with the stationary phase with the same strength. Thus, the retention of the *n*-alkanes increases in a predictable and regular manner.

The free energy of partitioning is related to measures of retention through the distribution constant as follows:

$$\Delta G° = -RT \ln K = -RT \ln(k\beta) = -2.303RT(\log k + \log \beta) \qquad (2.23)$$

where $R$ is the universal gas constant, $T$ is the temperature, $K$ is the distribution constant, $k$ is the retention factor, and $\beta$ is the phase ratio, which is constant for any individual column. The last relationship in Equation 2.23 shows that because $\Delta G°$ is linearly related to the number of carbon atoms in an *n*-alkane, then $\log k$ is also linearly related to the number of carbon atoms. Indeed, plots of $\log k$ versus *n*-alkane carbon number are linear when measured on the same column held at a constant temperature. *Thus, the n-alkanes can act as regularly spaced pillars or benchmarks against which the retention of other compounds (non-n-alkanes) can be measured.* Note that it is the *logarithm* of the retention factor, and not the retention factor itself, that is linear with carbon number for the *n*-alkanes. This arises because of the logarithmic relationship between $\Delta G°$ and $k$. We also point out that the linearity between $\log k$ and $n_{CH_2}$ holds rather nicely for many other homologous series if one disregards the first few members of the series. Thus, a plot of $\log k$ versus $n_{CH_2}$ for the *n*-alkanols (*n*-propanol, *n*-butanol, etc.) forms a straight line with nearly the same slope as that for the *n*-alkanes. In all cases, such a plot yields a slope which, when multiplied by $-2.303RT$, yields the free energy of partitioning of a methylene unit.

Using the *n*-alkanes as benchmarks by which to measure the retention of other solutes, Kovats defined a retention index, *RI*, shown in Equation 2.24.[24]

$$RI = 100 \left( \frac{\log k_s - \log k_n}{\log k_{n+1} - \log k_n} \right) + 100(n) \tag{2.24}$$

The Kovats retention index compares the retention factor of a solute ($k_s$) to the two *n*-alkanes that elute just before ($k_n$) and after ($k_{n+1}$) it. Using this definition, the Kovats retention index of an *n*-alkane at any temperature on a column of any dimension is exactly its number of carbon atoms times 100. For example, at any temperature on a column of any dimension, the Kovats retention index for *n*-pentane is exactly 500.

The denominator in the first term of Equation 2.24 is related to the free energy of partitioning contributed by a methylene group, while the numerator indicates how close or far away the solute is from the less retained *n*-alkane. The entire first term, therefore, can be viewed as a fraction that measures how far along the free energy scale established by the linearity of the *n*-alkanes the solute is. This is more easily understood by considering the simulated chromatogram in Figure 2.23. Based on the chromatogram, taking the first peak as that arising from an unretained species (methane is often used in practice), the retention factors in Table 2.7 are obtained. Plugging in the retention factors for the solutes and alkanes yields a retention index of 655.4 for solute 1 and 781.9 for solute 2. By extension, for example, it can be seen that the Kovats retention index for a solute that elutes between undecane ($n = 11$) and dodecane ($n = 12$) has a retention index between 1100 and 1200. The closer it elutes to undecane, the closer its Kovats retention index is to 1100 and the closer it elutes to dodecane, the closer its retention index is to 1200.

So the Kovats retention index serves as a measure of solute retention relative to the retention of *n*-alkanes on the same type of stationary phase. However, it was stated that the virtue of the Kovats retention index is that it is independent of column length, mobile phase flow rate, *and stationary phase thickness* (in contrast to the retention factor, *k*, which depends on film thickness). The independence from column length and mobile phase velocity is clear in that the retention index is defined solely in terms of retention factors, which, as was already discussed, are independent of column length and mobile phase velocity. To understand the independence of stationary phase thickness, substitute the relationship

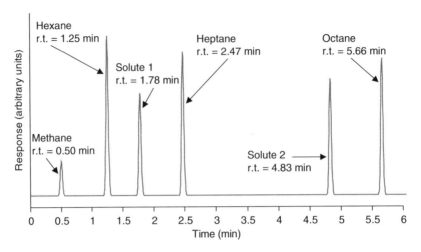

**FIGURE 2.23** Simulated chromatogram to illustrate Kovats retention index calculations.

**TABLE 2.7   Kovats Retention Indices Based on Figure 2.23**

| Solute | Retention time (min) | $k$ | $\log k$ | Retention index |
|---|---|---|---|---|
| Hexane | 1.25 | 1.50 | 0.176 | 600.0 |
| Heptane | 2.47 | 3.94 | 0.595 | 700.0 |
| Octane | 5.66 | 10.32 | 1.014 | 800.0 |
| Solute 1 | 1.78 | 2.56 | 0.408 | 655.4 |
| Solute 2 | 4.83 | 8.66 | 0.938 | 781.9 |

$k = K/\beta$ into the definition. Focusing only on the term in brackets yields

$$\left( \frac{\log k_s - \log k_n}{\log k_{n+1} - \log k_n} \right) = \left( \frac{\log \dfrac{K_s}{\beta} - \log \dfrac{K_n}{\beta}}{\log \dfrac{K_{n+1}}{\beta} - \log \dfrac{K_n}{\beta}} \right) = \left( \frac{(\log K_s - \log \beta) - (\log K_n - \log \beta)}{(\log K_{n+1} - \log \beta) - (\log K_n - \log \beta)} \right)$$

(2.25)

Because $\beta$ does not depend on the solute but rather is a constant for any individual column, $\log \beta$ subtracts out in the numerator and denominator, leaving the result

$$\mathrm{RI} - 100 \left( \frac{\log K_s - \log K_n}{\log K_{n+1} - \log K_n} \right) + 100(n)$$

(2.26)

Fundamentally, the distribution constant for any one specific type of stationary phase depends solely on the temperature. Thus, the Kovats retention index of a solute on a particular type of stationary phase is temperature dependent but is independent of phase ratio, column length, and mobile phase velocity, making it one of the most reliable measures of retention when comparing different GC systems or results from different laboratories.

**EXAMPLE 2.10**

Determine the Kovats retention index for a solute with a retention time of 8.72 min if octane elutes at 8.13 min, nonane at 9.86 min, and methane at 0.72 min.

**Answer:**
First determine the retention factors for octane (C8), nonane (C9), and the solute (S) using methane as the unretained (dead time) marker.

$$k_{C8} = \frac{t_r - t_m}{t_m} = \frac{8.13 - 0.72}{0.72} = 10.29$$

$$k_{C9} = \frac{t_r - t_m}{t_m} = \frac{9.86 - 0.72}{0.72} = 12.69$$

$$k_S = \frac{t_r - t_m}{t_m} = \frac{8.72 - 0.72}{0.72} = 11.11$$

Then apply the Kovats retention index equation:

$$RI = 100 \left( \frac{\log k_s - \log k_n}{\log k_{n+1} - \log k_n} \right) + 100(n) = 100 \left( \frac{\log k_s - \log k_{C8}}{\log k_{C9} - \log k_{C8}} \right) + 100(8)$$

$$RI = 100 \left( \frac{\log 11.11 - \log 10.29}{\log 12.69 - \log 10.29} \right) + 100(n) = 100 \left( \frac{0.03330}{0.09104} \right) + 800 = 836.6$$

Another problem: Determine the Kovats retention index for a solute with a retention time of 5.89 min if pentane elutes at 4.61 min, hexane at 6.03 min, and methane at 0.51 min.

**Answer:** 591.4.

The retention index has also been used to compare the chemical nature of different stationary phases.[25,27] For example, on a DB-5 column (5% diphenyl 95% dimethyl polysiloxane in Figure 2.6), phenol has a retention index of 973. On a DB-1301 column (6% cyanopropylphenyl 94% dimethyl polysiloxane in Figure 2.6), it has a retention index of 1119. From this, one can conclude that the DB-1301 column is more polar and possibly a better hydrogen bond acceptor than a DB-5 column. Comparing the structures of these two stationary phases (see Figure 2.6) shows that this analysis is consistent with the structures, and that the retention index indeed reflects the different natures of stationary phases. This is important because being able to probe differences in stationary phases leads to more insightful, productive, and judicious selection of columns for complex separations and analyses.

***2.4.1.1. Retention Indices and Intermolecular Interactions.*** A comparison of Kovats retention indices of some select solutes measured on different columns is quite revealing regarding the influence of intermolecular interactions on retention and ultimately on the selection of columns for particular separations. For example, consider retention on a

**TABLE 2.8 Kovats Retention Indices at 80 °C**

| Column[a] | Acetone | DMSO | Nitropropane | Hexafluoroisopropanol | Benzene | Phenol | 2-Pentanone | Butylether | Ethanol |
|---|---|---|---|---|---|---|---|---|---|
| DB-1 | 473 | 791 | 704 | 471 | 664 | 951 | 663 | 880 | 423 |
| DB-17 | 607 | 1093 | 911 | 465 | 791 | 1160 | 802 | 931 | 536 |
| DB-210 | 808 | 1400 | 1114 | 649 | 800 | 1179 | 990 | 929 | 586 |
| DB-225 | 731 | 1435 | 1126 | 1007 | 842 | 1621 | 941 | 943 | 697 |
| DB-wax | 846 | 1573 | 1230 | 1399 | 979 | _[b] | 1008 | 977 | 955 |

[a]Shown in increasing polarity.
[b]Not available due to excessive retention.

116

DB-1 column (100% dimethyl polysiloxane). The retention of all solutes on this stationary phase is generally dominated by dispersion interactions, regardless of the polarity of the solute. The Kovats retention indices for a number of solutes on this stationary phase are shown in Table 2.8 and plotted in Figure 2.24. On this column, acetone elutes between butane and pentane as indicated by its Kovats retention index being between 400 and 500. When considering the small size and relatively high volatility of acetone, it is not surprising that it is not well retained on a column dominated by dispersion interactions. However, on a polar column such as DB-225 (50% cyanopropyl 50% phenylmethyl polysiloxane) or DB-wax (polyethyleneglycol), the retention of acetone increases and it elutes closer to the less volatile *n*-alkanes, heptane, and octane. This increase in retention is due to dipole–dipole interactions between acetone and the stationary phase that exist on the DB-225 and wax phases that are absent on the nonpolar DB-1 phase. Even the DB-17 column (50% phenylmethyl polysiloxane) increases the retention of acetone relative to the *n*-alkanes. This increase in retention arises in part from the dipole of acetone inducing a dipole in the polarizable phenyl rings in the stationary phase as depicted in Figure 2.25.

The differences in solute retention on nonpolar and polar columns are even more dramatic for solutes capable of donating hydrogen bonds. For example, hexafluoroisopropanol elutes between butane and pentane on a DB-1 column (100% dimethyl polysiloxane) but elutes close to decane on a DB-225 phase (50% cyanopropyl 50% methylphenyl polysiloxane) and close to tetradecane on the wax phase! These are very dramatic increases in retention. They arise from the fact that hexafluoroisopropanol is

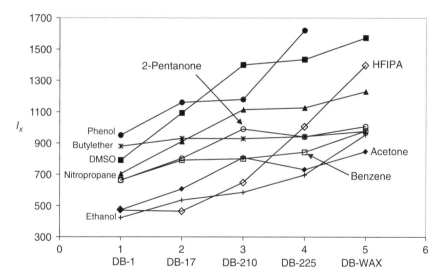

**FIGURE 2.24** Kovats retention indices of select compounds on five stationary phases at 80°C. Crossovers indicate changes in the elution order of solutes as a result of the different blends of intermolecular interactions the solutes experience with each stationary phase. Solutes: (♦) acetone, (■) dimethylsulfoxide (DMSO), (▲) nitropropane, (◊) hexafluoroisopropanol (HFIPA), (□) benzene, (●) phenol, (○) 2-pentanone, (*) butylether, (+) ethanol. Stationary phases: DB-1 = nonpolar dimethylpolysiloxane, DB-17 = intermediate polarity (14% cyanopropylphenyl)-methylpolysiloxane, DB-210 = selective for lone-pair electrons (50% trifluoropropyl)-methylpolysiloxane, DB-225 = polar (50% cyanopropylphenyl)-methylpolysiloxane, DB-wax = polar polyethylene glycol (PEG).

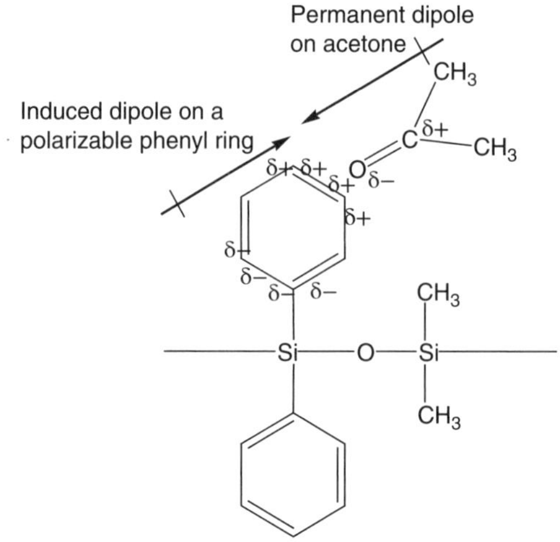

**FIGURE 2.25** Depiction of the permanent dipole of acetone inducing a dipole on the aromatic phenyl rings in a methylphenyl polysiloxane stationary phase.

a very good hydrogen bond donor and can therefore interact with the hydrogen bond accepting sites on the DB-225 cyanopropylphenyl-methylpolysiloxane stationary phase, whereas the alkanes, which form the basis for the Kovats retention index system, cannot. Similar effects are evident for other hydrogen bond donating solutes such as phenol, *m*-cresol, and ethanol.

**EXAMPLE 2.11**

1. Draw the structure of hexafluoroisopropanol.
2. Draw a picture showing it donating a hydrogen bond to another molecule such as butanone.
3. Based on the structure and electron-withdrawing arguments, explain why hexafluoroisopropanol is a very strong hydrogen bond donor and has very little hydrogen-bond-accepting ability.

**Answer:**

1.

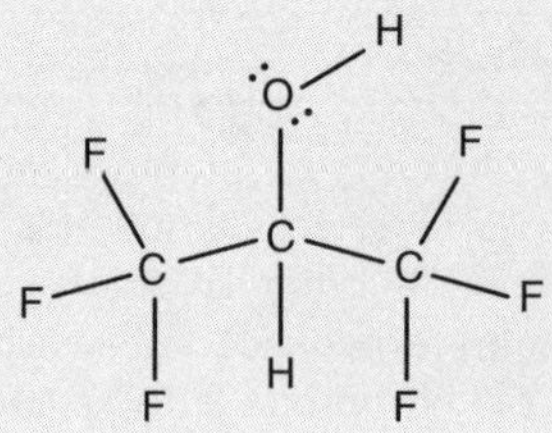

2.

3. Because fluorine atoms are highly electronegative, they pull electron density away from the C—O—H bonds, meaning that there is even less electron density around the hydrogen atom compared to regular isopropanol. This means the partial positive charge on the hydrogen atom in hexafluoroisopropanol is even greater compared to isopropanol, and thus participates in stronger hydrogen bond interactions with other molecules. By the same reasoning, hexafluoroisopropanol is a poor hydrogen bond acceptor because the lone pairs on the oxygen, which are normally responsible for accepting hydrogen bonds, are pulled toward the electronegative fluorine atoms. This reduces the electron density around the oxygen atom, making it a very weak hydrogen bond acceptor.

### 2.4.1.2. *Kovats Retention Indices and Column Properties.* Note that as shown in Figure 2.24, polar solutes such as nitropropane and 2-pentanone have significantly increased retention on polar columns like DB-225 (50% cyanopropyl 50% methylphenyl polysiloxane) and DB-wax (polyethylene glycol, PEG) relative to their retention on DB-1 columns (100% dimethyl polysiloxane), but the change is not as significant as the hydrogen-bond-donating solutes discussed above. This shows that the DB-225 and wax columns are indeed capable of accepting hydrogen bonds as their structures suggest, and that these interactions work in concert with dispersion and dipole–dipole interactions to dictate overall solute retention. It is worth noting that the retention of butylether, which has only a very small dipole moment (1.13 Debye) and is therefore retained mainly by dispersion interactions, is nearly the same on all of the columns regardless of the column structure. *This indicates that while a column may be capable of interacting through a suite of interactions (e.g., dispersion, dipole–dipole, and hydrogen bonding), that capability is only important if the solutes have* **complementary properties** *that make them capable of participating in these interactions with the stationary phase.*

### 2.4.2. Stationary Phase Selection

Perhaps the most important concept to learn from the data in Table 2.8 and Figure 2.24 is that the nature of the stationary phase can so profoundly influence solute retention that *the order of elution can be different on different columns*. For example, hexafluoroisopropanol elutes well *before* 2-pentanone on columns that generally retain compounds through dispersion interactions such as DB-1 (100% dimethyl polysiloxane) and DB-17 (50% diphenyl 50% dimethyl polysiloxane), but it elutes *after* 2-pentanone on columns capable of accepting hydrogen bonds like DB-225 (50% cyanopropyl 50% methylphenyl polysiloxane) and DB-wax (polyethylene glycol). In contrast, the retention index of butylether is quite insensitive to the nature of the stationary phase. Many other examples of changes in elution order can be found in Figure 2.24. They arise because each solute/stationary phase pair is governed by different types and strengths of intermolecular interactions. It is clear, then, that column choice can significantly influence the separation factors, and thus resolution, when a wide variety of solutes is present.

Further analysis of Figure 2.24 shows that the overall elution order of all of the compounds is different on each phase because the retention indices change dramatically relative to one another from column to column. In addition, imagine if this solute set is in fact a mixture of interest and that 2-pentanone is the solute that one wants to quantify. The plot makes it clear that 2-pentanone is most clearly resolved from other compounds (i.e., has a distinct Kovats retention index) on the DB-210 phase (50% trifluoropropyl 50% dimethyl polysiloxane). In fact, on that phase, only two of the solutes nearly coelute (acetone and benzene), whereas multiple compounds have nearly identical retention indices on many of the other stationary phases. Thus, if the components in a particular sample mixture cannot be separated on one column, a different column may be necessary to achieve the desired separation. In this way, information such as that found in Figure 2.24 can guide stationary phase selection to optimize resolution, and thus quantitation, of an analyte.

**EXAMPLE 2.12**

Referring to Figure 2.24, what is the elution order of the nine solutes on a DB-1 phase? On a DB-210 phase?

| Elution order | DB-1 | DB-210 |
|---|---|---|
| 1 | Ethanol | Ethanol |
| 2 | Acetone = HFIPA | HFIPA |
| 3 | | Acetone = benzene |
| 4 | Benzene = 2-pentanone | |
| 5 | | Butylether |
| 6 | Nitropropane | 2-Pentanone |
| 7 | DMSO | Nitropropane |
| 8 | Butylether | Phenol |
| 9 | Phenol | DMSO |

Note that there are a number of elution order changes, and dimethylsulfoxide experiences a significant increase in Kovats index on the DB-210 phase, consistent with the information that the DB-210 phase is selective for lone-pair electrons.

**Another question:**
By referring to the figure of stationary phases (Fig. 2.6) on page 74, suggest a column that could be used for the analysis of nitrobenzene, diethyl ether, 2-heptanone, aniline, and 1,2-dichlorobenzene.

**Answer:**
These are relatively polar compounds, so any of the intermediately polar or polar stationary phases, such as Rtx-50, Rxi-17, or Rtx-1701, would be a logical selection. Polar or very polar columns may yield too much retention and therefore result in unnecessarily long retention times, so the first choice would be one of the intermediately polar compounds.

While this section focuses on stationary phase selection, it must be remembered that resolution problems can often be overcome by changing the operating temperature or the temperature program. Recall that temperature has a dramatic effect on retention and the separation factor. Furthermore, changing the oven temperature is considerably easier than changing the column, which is time-consuming. In practice, then, it is often more practical to vary the operating conditions (temperature, flow rate, etc.) before changing the column.

When attempting analyses for the first time, especially when the polarities of the solutes are truly unknown or span a significant range, columns such as DB-5 (5% diphenyl 95% dimethyl polysiloxane) or DB-35 (35% diphenyl 65% dimethyl polysiloxane) provide reasonable separation factors for both polar and nonpolar solutes. Thus, these columns, along with DB-1 columns (100% dimethyl polysiloxane), tend to be the workhorses for routine GC analyses, with the other columns employed for more specific applications.

One final consideration regarding stationary phase selection is that the polarity of the column should generally match the polarity of the compounds being separated. When this criterion is not met, poor peak shapes can result from the fact that the solute solubility in the stationary phase may be quite low. If the solubility is low, the column is easily overloaded with solute. This results in adsorption of the solute *onto the surface* of the stationary phase in addition to true absorption (partitioning) *into* the stationary phase (see Figure 2.26).[20] This mixed mode of retention results in poor peak shapes. This effect can be particularly pronounced with amines and carboxylic acids analyzed on nonpolar stationary phases. In addition to adsorption onto the stationary phase, polar solutes can also be adsorbed onto the surface of glass injection port liners and to bare silanol (—Si—O—H) groups in the column. This also leads to distorted peak shapes. It is therefore important to use liners and columns that have been "deactivated," meaning that steps have been taken to chemically modify residual silanol groups to convert them into nonpolar entities.

### 2.4.2.1. McReynolds Constants as a Guide for Stationary Phase Selection.
As described in the earlier section, Kovats retention indices provide information about the types and strength of intermolecular interactions between solutes and stationary

ABsorption                    ADsorption
(a)                           (b)

**FIGURE 2.26** Cartoon illustrating the difference between absorption into the stationary phase and adsorption onto the surface. (*Source:* James M. Miller, *Chromatography: Concepts and Contrasts*, John Wiley and Sons, Inc., 2005, page 45. Reprinted with permission.)

phases. McReynolds and Rohrschneider used this fact to characterize stationary phase liquids.[25-27] They selected a small number of representative test compounds with different functional groups and measured their Kovats retention indices on over 200 phases. The test compounds were chosen to represent specific types of intermolecular interactions. The commonly cited test compounds are listed in Table 2.9A. Squalane, which is a nonpolar, aliphatic phase and thus interacts with solutes via dispersion and dipole–induced dipole interactions only, was selected as a reference stationary phase. A quantity, $\Delta I$, was defined where

$$\Delta I = I_{\text{liquid phase}} - I_{\text{squalane}} \tag{2.27}$$

and $I_{\text{liquid phase}}$ and $I_{\text{squalane}}$ are Kovats indices of the reference compounds on columns coated with the liquid of interest and squalane, respectively. Polar, hydrogen-bonding compounds like $n$-butanol and 2-pentanone are more strongly retained on polar phases capable of hydrogen bonding than they are on squalane, leading to positive $\Delta I$ values. Larger $\Delta I$ values therefore signal larger differences in the strength of specific intermolecular interactions for the phases of interest relative to squalane. Furthermore, because each test solute was judiciously selected to probe particular intermolecular interactions, the $\Delta I$ values for each solute reveal specific differences between the stationary phase liquids in terms of their interaction ability.

The symbols $X'$, $Y'$, $Z'$, $U'$, and $S'$ were selected to represent $\Delta I$ values obtained with the five different probe solutes shown in Table 2.9A. The values of $X'$, $Y'$, $Z'$, $U'$, and $S'$ are known as McReynolds constants or McReynolds numbers. Numeric values for a few common liquid stationary phases are shown in Table 2.9B. The last column in Table 2.9B shows the sum of the individual values, with greater sums generally being associated with

**TABLE 2.9A   McReynolds Probe Solutes, Parameter Symbols, and Interactions They Represent**

| Probe solutes | Parameter symbol | Stationary phase liquid property probed |
|---|---|---|
| Benzene | $X'$ | Ability to interact with polarizability arising from aromaticity and unsaturated bonds |
| n-Butanol | $Y'$ | Ability to interact by donating and accepting hydrogen bonds |
| 2-Pentanone | $Z'$ | Ability to interact by donating hydrogen bonds |
| Nitropropane | $U'$ | Ability to retain solutes via dipole–dipole interactions |
| Pyridine | $S'$ | Ability to interact with strong proton-accepting solutes |

**TABLE 2.9B   McReynolds Constants for Common GC Stationary Phases[7]**

| Liquid stationary phases | $X'$ | $Y'$ | $Z'$ | $U'$ | $S'$ | Sum |
|---|---|---|---|---|---|---|
| Squalane | 0 | 0 | 0 | 0 | 0 | 0 |
| Dimethyl siloxane (OV-1) | 15 | 53 | 44 | 64 | 41 | 219 |
| 10% Diphenyl 90% dimethyl polysiloxane (OV-3) | 44 | 86 | 81 | 124 | 88 | 423 |
| 14% Cyanopropylphenyl 86% dimethyl polysiloxane (OV-1701) | 67 | 170 | 153 | 228 | 171 | 789 |
| 50% Diphenyl 50% dimethyl siloxane (OV-17) | 119 | 158 | 162 | 243 | 202 | 884 |
| 50% Trifluoropropyl 50% dimethyl polysiloxane (OV-210) | 146 | 238 | 358 | 468 | 310 | 1520 |
| Carbowax 20M (polyethylene glycol) | 322 | 536 | 368 | 572 | 510 | 2308 |
| Diethylene glycol succinate (DEGS) | 496 | 746 | 590 | 837 | 835 | 3504 |

increasing polarity of the liquid phases. However, it should be noted that simply looking at the sum risks missing important information about specific intermolecular interactions present in the individual parameters.

Comparing a few of the McReynolds numbers shows how this type of column characterization is useful. For example, compare the values for 14% cyanopropylphenyl 86% dimethyl polysiloxane to those of 50% diphenyl 50% dimethyl siloxane. Doing so shows that the biggest difference between these columns is the $X'$ parameter, which represents the phases' susceptibility to interactions arising from solute polarizability. It can be concluded from these numbers that of the two liquid phases, the 50% diphenyl 50% dimethyl siloxane phase offers greater interaction, and therefore potential selectivity, for aromatic species. This could help select a column for separating a mixture in which many of the compounds are known to have aromaticity or unsaturated bonds. The increased interactions with polarizable compounds arise because of the high density of phenyl rings in the 50% diphenyl 50% dimethyl phase. In a similar manner, comparing Carbowax 20M (polyethylene glycol, PEG) to diethylene glycol succinate (DEGS) shows that both are extremely polar relative to squalane (i.e., have high sums), and that of the two, DEGS has stronger interactions with all classes of solutes (although both are commonly thought of as extremely retentive). This increased interaction of DEGS relative to PEG arises from the modification of the polyethylene glycol backbone, which introduces polar, hydrogen-bonding ketone moieties to the stationary phase. Therefore, DEGS offers greater potential retention and selectivity for a mixture of polar, hydrogen-bonding solutes than does Carbowax 20M. These examples show that matching the solutes' properties to the column characteristics as described by the McReynolds constants can inform the selection of stationary phases.

## 2.5.  GAS CHROMATOGRAPHY IN PRACTICE

The previous sections dealt with the theory and instrumentation of GC, specifically focusing on the intermolecular interactions that control retention and how instruments are designed to introduce, separate, and detect components in mixtures. In addition to these fundamental principles, there are several practical aspects to keep in mind when performing GC analyses. These are discussed in the following sections.

### 2.5.1.  Syringe Washing

An important aspect of the practice of gas chromatography is syringe washing. Because the same syringe is used repeatedly for different samples, it is important to rinse the syringe multiple times between analyses, especially when different samples are being used. The reason for this is that some sample solution remains on the inside walls of the syringe after an injection. If the syringe is then used for a different sample, components from the first sample now contaminate the second sample. They will be injected, produce peaks, and appear as if they really are in the second sample. This is referred to as carryover. To eliminate carryover, the syringe is flushed with solvent multiple times by drawing solvent into the syringe and then expelling it into a waste vial. Once this is done, the syringe now has solvent coating the walls. If sample is drawn up, the additional solvent in the syringe dilutes the sample by a small factor. This would lead to inaccurate quantification of the components. For this reason, sample is drawn up into the syringe and expelled into a waste vial multiple times prior to injection in a manner similar to conditioning a pipet or buret. It is poor technique to draw the sample and expel it back into the sample vial because this can lead to contamination from residual solutes in the syringe. While rinsing the syringe with solvent and sample can be done manually, instruments with autosamplers can be programmed to conduct these steps as part of the injection process.

### 2.5.2.  Controls and Blanks/Ghost Peaks

In every analytical method, it is necessary to analyze control and blank samples.[28] In GC, the blank analysis entails injecting the solvent in which the sample is dissolved. Most solvents are not pure and therefore produce more than a single peak. If the blank is not performed, the analyst may *incorrectly* identify peaks due to impurities in the solvent as components of the sample. It is therefore wise to periodically inject just the solvent during a trial in which multiple samples are analyzed. It is also good practice to periodically inject control samples containing known compounds, typically the analytes, to check the reproducibility of retention times and peak areas. Another benefit to analyzing controls and blanks is that the analyst may detect "ghost peaks." These arise from solutes that are so well retained that they do not elute during the course of the first sample injection but instead elute sometime after a subsequent injection has started, making it look as though they were part of that later sample when really they were part of the first sample. They tend to not have reproducible retention times because with each new injection, the recording of time is started over. Because the compound giving rise to the ghost peak is already part way through the column from the previous separation, it was not injected with the other solutes and will therefore seem to "wander" in terms of its retention time. Because the pure solvent should produce a relatively simple and repeatable chromatogram, periodically injecting just the solvent can help detect the presence of ghost peaks. To eliminate ghost peaks, the final column temperature can be increased to decrease the retention of the solute causing the ghost peak. In addition, the hold time at the final temperature can be extended to ensure that all solutes have eluted before starting another injection.

### 2.5.3. Autosamplers

As stated earlier, sample injections are made either manually or using an autosampler. An autosampler typically consists of a syringe, a tray that allows multiple sample vials to be housed near the injection port, and a robotic arm. The robotic arm selects the sample to be analyzed and places it in a position where the syringe can withdraw a sample. The syringe then rapidly and reproducibly injects the sample into the injection port. Because they are automated, autosamplers make it possible for hundreds of injections to be made sequentially without an operator present. They also offer superior precision over manual injections. For these reasons, they are widely used throughout industry to increase the efficiency and precision of analyses. Regardless of which method is used, the syringe washing and carryover effects discussed above remain the same.

Despite their time-saving and improved efficiency, autosamplers are not infallible and their use can result in bias. One common form of bias results from the fact that the sample trays are often located close to the heated injection port. Because of this proximity, the heat from the injection port can cause the solvent in the sample vials to evaporate over extended periods of time, even if they are sealed. Because many autosampler programs are set to run for multiple hours or days, the evaporation can serve to concentrate the less-volatile solutes. If this occurs, when these samples are analyzed, the solute concentrations will be artificially high, producing inaccurate results. These effects, however, can be checked for simply by injecting the same sample at the start, middle, and end of an entire series of analyses. If the peak areas and heights remain the same, then no evaporation has occurred. If they systematically increase with time, some bias is present and the results are not to be trusted.

### 2.5.4. GC Septa

The syringe is rapidly pressed through a septum made of silicone rubber to inject the sample into the injection port. Once the syringe is removed, the flexible septum material seals off the hole made by the needle. After repeated injections, the needle eventually bores a hole that cannot be sealed, resulting in leaks through the septum. This affects retention time reproducibility and quantitative analyses because both the carrier gas and injected solutes can escape from the system via the hole in the septum. It is therefore important to check the integrity of the injection port septum and replace it as needed to prevent adverse chromatographic results.

### 2.5.5. Qualitative Analysis

Reproducible measures of retention are important in GC for solute identification purposes. Many common detectors such as flame ionization and thermal conductivity detectors do not yield solute-specific information other than the retention time. Thus, to identify a compound in a chromatogram of a complex mixture, it is necessary to inject known compounds. If the unknown solute and known compound have identical measures of retention (e.g., Kovats RI), then it is possible that the solute is indeed the same compound as the standard. If the two compounds differ in retention, then they are not the same compound. Thus, in order to identify a solute, accurate and precise measures of retention are required. Naturally, using detectors that also provide structural information, such as infrared and mass spectrometry detectors, significantly help to identify compounds.

### 2.5.6. Quantitative Analysis

While GC can be used to identify unknown compounds, it is more often used to *quantify* solutes that are known to be in a sample. This is accomplished by preparing calibration curves of known standards and measuring either peak area or peak height. When using GC for quantitative purposes, using calibration curves based on internal standards (IS) is the best practice. The reason for this is that internal standards can compensate for undesirable variability that occurs in GC systems, especially during the chaotic and somewhat irreproducible injection process.

***2.5.6.1. Quantitative Analysis: Internal Standards.*** In a typical GC analysis, $1\,\mu L$ of sample is injected. It is difficult to *reproducibly* draw up and deliver such a small volume of sample, even for autosamplers. Any variation in this amount manifests itself in a comparable variation in peak area or height. Intentionally including an internal standard in calibration solutions as well as in the sample can reduce this variation. Consider the following. Suppose you want to quantify the exact amount of DEET (*N,N*-diethyl-*m*-toluamide) in insect repellant that contains approximately 22% DEET. One could make five standards, containing 0.5%, 1%, 2%, 3%, and 4% DEET in a solvent such as methanol. The repellant would be diluted in the same solvent by a factor of 1:10 to get the sample in range with the standards in a dilute enough concentration so as to yield good chromatographic peaks.

Suppose that exactly $1.000\,\mu L$ of the 0.5% sample is actually injected (as unlikely as this is) and produces a peak area of 400 arbitrary units. One could then attempt to inject *exactly* $1.000\,\mu L$ of the rest of the standards and the sample. But again, it is difficult to inject the exact same amount each time. So suppose the amounts shown in Table 2.10 are the amounts that *actually* get injected. Keep in mind that the analyst would not know these exact differences, thinking instead that they had actually injected $1\,\mu L$ each time. (It should also be noted that the variations in the table are slightly exaggerated compared to actual reproducibilities to make the effect clearer.)

TABLE 2.10   Illustration of the Effects of Injection Variation on Peak Area

| DEET (%) | Amount injected (μL) | Expected DEET area (arbitrary units) | Actual DEET peak area (arbitrary units) |
|---|---|---|---|
| 0.5 | 1.00 | 400 | 400 |
| 1 | 0.85 | 800 | 680 |
| 2 | 1.10 | 1600 | 1760 |
| 3 | 1.05 | 2400 | 2520 |
| 4 | 0.92 | 3200 | 2944 |

TABLE 2.11   Utility of Internal Standards for Quantitative Analysis

| % DEET | Amount injected (μL) | DEET peak area (arbitrary units) | I.S. peak area (arbitrary units) | DEET/I.S. ratio |
|---|---|---|---|---|
| 0.5 | 1.00 | 400 | 1000 | 0.40 |
| 1 | 0.85 | 680 | 850 | 0.80 |
| 2 | 1.10 | 1760 | 1100 | 1.60 |
| 3 | 1.05 | 2520 | 1050 | 2.40 |
| 4 | 0.92 | 2944 | 920 | 3.20 |

Based on the first 0.5% sample in which 1.000 μL was actually delivered, a perfect calibration curve would require a perfect doubling of the peak area (i.e., 800 arbitrary units) for the 1.00% sample. If, however, 0.85 μL is injected instead, a peak area of only 800(0.85) = 680 arbitrary units results. Likewise, down the list, the variation in the amount injected causes deviations from the perfect calibration curve that would result if exactly 1.000 μL is injected each time.

Suppose now that, instead of just the DEET, an internal standard such as *N,N*-dimethyl-*m*-toluamide is also added at a concentration of 2% in *all* of the standards and samples. In this case, the calibration curve is a plot of the *ratio* of the peak area of the compound of interest (DEET) relative to the internal standard.

Suppose the same variation in injection volume occurs as was detailed above. How does the internal standard help? *The answer is that anything that happens to the DEET also happens to the internal standard.* Let's assume that exactly 1.000 μL of solution containing 2% of the internal standard produces a peak area of 1000 arbitrary units for the internal standard. Consider Table 2.11 and compare it to Table 2.10.

Based on the ratio of 0.40 that results from the first analysis, the second calibrant that contains twice as much DEET but the same amount of internal standard should have a ratio of 0.80. We saw that if only 0.85 μL are injected, the DEET peak area falls short of the expected value of 800. However, less of the internal standard is also injected. Furthermore, the *percentage* decrease in the amount injected is the same for the internal standard as the solute, so the internal standard peak area is only 850 compared to the expected 1000, but the *ratio* of DEET/internal standard is 680/850 = 0.800, *which is exactly what was expected based on the ratio from the first sample.* Thus, the use of internal standards greatly reduces the *variation* in the calibration curve that results from variation in the injection process. A statistical analysis and visual inspection of the two resulting calibration curves (with and without the internal standard) would show that the analysis performed with the internal standard is analytically superior to the one without.

It is necessary to emphasize that the discussion above assumes accurate preparation of the standard solutions in such a manner that the internal standard has the same concentration in each solution. Also, note the structural similarity of the internal standard (*dimethyl* toluamide) and the compound of interest (*diethyl* toluamide). Matching the internal standard to the analyte helps ensure that whatever happens to the analyte happens to the same extent to the internal standard.

---

**EXAMPLE** *2.13*

Internal standards are primarily used to mitigate the effects of what problem in GC?

**Answer:**
Internal standards mitigate the effects caused by variations in the amount of solution injected and by variations in the injection/vaporization process itself.

---

### 2.5.7.  Derivatization

Some compounds, particularly those containing amine and carboxylic acid functional groups, sometimes produce poor peak shapes or are not volatile enough to easily analyze with GC. In these cases, the compounds can be derivatized prior to analysis using common derivatization schemes. Modification of these polar functional groups reduces their intermolecular interactions with any bare fused silica not covered with stationary phase. Because these interactions with the wall are one of the main sources of poor peak shapes, derivatization improves the peak shape and makes it easier to quantify the area under the peak. Many environmental and pharmaceutical compounds contain amines, so these techniques are particularly useful in those fields. In addition, compounds can be derivatized with fluorocarbon tags. When used in combination with electron capture detectors, these tags can significantly decrease detection limits and increase the separation factor of the methods.[30] Common derivatization reagents are detailed elsewhere.[29-33]

### 2.5.8.  High-Speed GC

A trend in the practice of gas chromatography has been toward faster separations – separations that occur within 2 minutes – in contrast to the 5–60 minute separation times that are currently quite common. This is known as high-speed gas chromatography (HSGC).[34-38]

Such separations require that all sources of band broadening be greatly reduced so that all of the components in a given mixture produce sharp, narrow peaks that can be resolved from one another in short periods of time. Technologies that allow small vapor plugs, highly efficient narrow-bore capillary columns, rapid temperature programming, and fast detection techniques are available and make HSGC possible. High-speed separations are performed on shorter columns at higher flow rates than conventional separations. Both of these changes negatively impact the number of theoretical plates on a column, and thus can adversely affect resolution. To compensate for this loss in plates, narrow-bore columns with thin films are used. These columns tend to be more efficient than thicker films and thus partially offset the loss in plates caused by the use of shorter columns.

The interest in high-speed analysis is driven by the fact that time is money. The more analyses performed per unit time, the more efficient is the process. More analyses per unit time also allow more complex studies to be conducted and better real-time monitoring of continuous processes.

### 2.5.9. Tandem GC

Another development in gas chromatography is the tunable separation factors achieved by coupling two chemically distinct stationary phases in tandem and varying the amount of time a solute mixture resides in each.[38–43] A simple schematic of a tunable GC system is shown in Figure 2.27.[42] The use of a pressure controller at the junction $P_t$ allows the fraction of the total carrier gas pressure drop to be differentially applied over the two different columns. If the pressure at $P_t$ is very low compared to the inlet pressure ($P_1$), the carrier gas velocity in the first column is very high, meaning that solutes reside in column 1 for only a short period of time in comparison to the amount of time they spend in column 2. In this case, the total separation that is achieved is largely dictated by the separation achieved on the second column.

Conversely, if $P_t$ is close to $P_1$, column 1 has the dominant influence on the separation factor compared to column 2. Thus, by varying $P_t$, the relative influences of columns 1 and 2 can be varied to achieve the desired *overall* separation.

Tandem GC greatly facilitates high-speed GC because it is difficult to separate complex mixtures in a short amount of time using only the intermolecular interactions available on a single column. The use of two columns, however, provides increased discrimination power to achieve complex, high-speed separations. As this field advances, more techniques are being developed to take advantage of the differences in the two columns. Commercial instruments using high-speed tandem GC have potential applications in industrial, environmental, governmental, and academic laboratories.

### 2.5.10. Microfabricated GC

The need for high-speed separations and the desirability of continuous monitoring systems, combined with recent advances in microfabrication techniques has led to the creation of micro-GC systems.[7,43] In these systems, rectangular channels of very small dimensions are etched into a solid material, typically silicon. In Figure 2.28, the channels shown are 150 µm wide and 240 µm deep. The small dimensions of the channels combined with the precise control of the etching process allow for long columns to be created on very small silicon wafers or "chips." In this figure, the total column length is a remarkable 0.9 m despite the fact that it is etched into a wafer that is only 1.7 cm on a side. Common stationary

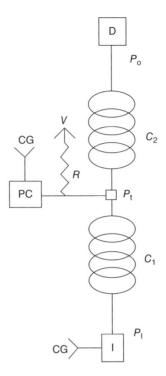

**FIGURE 2.27** Schematic of a tunable GC system. Notation: capillary columns with different stationary phases ($C_1$ and $C_2$), inlet (I), detector (D), electronic pressure controller (PC), carrier gas (CG) inlets, capillary pneumatic restrictor (R), vent point (V), inlet pressure ($P_I$), outlet pressure ($P_o$), and tuning pressure ($P_t$). $P_I$ is the highest pressure and $P_o$ is the lowest pressure. By varying the tuning pressure ($P_t$), the average carrier gas velocity in $C_1$ and $C_2$ is altered. Increasing $P_t$ decreases the pressure drop across $C_1$ and increases it across $C_2$. This decreases the average carrier gas velocity in $C_1$ and hence increases the residence time of the sample in $C_1$. Concomitantly, an increase in $P_t$ increases the carrier gas velocity in $C_2$, which decreases the solute residence time in that column. Thus, by varying $P_t$, the relative contributions of the two columns to the overall separation can be tuned to optimize the separation. (*Source:* Reprinted with permission from Sachs, R.; Coutant, C.; Grall, A; *Anal. Chem.* 72, 524A–533A, Copyright (2000), American Chemical Society.)

phases are used to modify the surface of the channels. In order for these systems to operate unattended for long periods of time and to eliminate the need for bulky high-pressure gas tanks, air from the surroundings is pulled through the columns by a vacuum applied at the end of the column. The use of ambient air as the carrier gas presents some challenges in that it causes generally wider peaks than either hydrogen or helium. It can also cause the breakdown of some stationary phases such as polyethylene glycol. Furthermore, particles in the air can clog the channels. In addition to these challenges, because these systems tend to be used for air monitoring where the airborne concentrations of most analytes tend to be low, preconcentration of the analytes is necessary. The need for preconcentration is heightened by the fact that the detectors used in these system tend to be less sensitive that common GC detectors.[7] Both the preconcentrators and detectors are discussed below.

Preconcentration of solutes is achieved by pulling the samples through a bed of adsorptive material such as carbon beads. The adsorptive bed is then rapidly heated using resistive heating (i.e., passing a current through the material). As the temperature of the

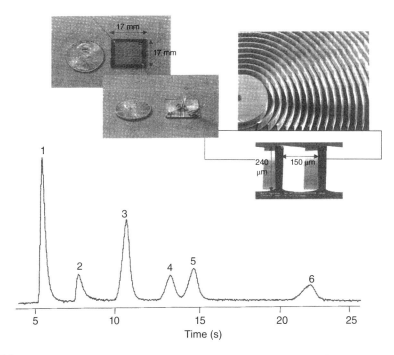

**FIGURE 2.28**  Micro-GC columns fabricated on silicon chips and a sample separation of six components in 25 seconds.  (*Source:* Reprinted from R.L. Grob and E.F. Barry, *Modern Practice of Gas Chromatography*, 4th edition, 2004, page 269 with permission from John Wiley and Sons, Inc.)

beads increases, the adsorbed solutes are released from the material and "injected" into the column, where they are separated according to their intermolecular interactions as described earlier in the chapter.

Flame ionization and thermal conductivity detectors – two of the most common detectors for GC – cannot be used for micro-GC systems because these detectors require external gas tanks and are difficult to miniaturize. Systems employing chemiresistor devices have been used for micro-GC systems. These devices contain gold particles coated with a monolayer of organic thiols (see Figure 2.29).[7,44] The effluent of the column passes over a thin layer of the particles. Organic solutes eluting from the column are adsorbed into the organic layer. This changes the electrical resistance of the gold particles. Thus, solutes are detected by the increase in the resistance they induce in the layer of gold particles. Different organic thiols respond differently to solutes (i.e., they have different selectivities). It is possible to not only detect but also identify components by incorporating several different types of organic thiols into the detector array and using pattern recognition methods to elucidate the chemical nature of the analytes.[7]

## 2.6.   A "REAL-WORLD" APPLICATION OF GAS CHROMATOGRAPHY

### 2.6.1.   GC and International Oil Trading

In April 2000, a crew from a U.S. Navy destroyer, the USS Russell, boarded a Russian tanker, the Akademik Pustovoit, which belonged to Royal Dutch Shell, on the suspicion that it

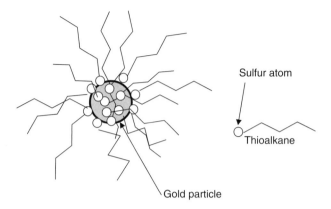

**FIGURE 2.29**   Gold particle coated with thiobutane.

was illegally carrying Iraqi oil in violation of sanctions established as a result of the 1991 Persian Gulf War.[45] The tanker was in international waters, having left Iranian territorial waters.

Samples of the oil were taken from the tanker and sent to a U.S. Customs Laboratory in San Francisco for analysis. The results of the tests are revealed later in this section, but first imagine the international incident that could arise if an incorrect analysis is performed. The United States, Russia, Iran, and Iraq were all involved in this incident. Mistakes could lead to heightened international conflict. The analysis had to be done correctly to avoid such a situation.

From an analytical chemistry standpoint, the question arises: "Can chemical analysis determine the country from which the oil originated?" Given that oil is a mixture of relatively volatile organic compounds, GC appeared to be a logical choice for performing such an analysis.

Oil is formed by the decay of organic materials, and the starting materials and conditions under which they decay vary with geographic location. Thus, oil from different regions can be identified by differences in its specific chemical composition. These differences are due to the way the original material was deposited in the soil, the conditions it experienced while decaying, the length of time it has decayed, and the various components that it mixed with throughout time. Using these different components, referred to as "biomarkers," the country of origin of an unknown sample can be determined by comparing chromatograms of known samples to the unknown sample, similar to the comparison of unknown fingerprints to known prints to determine the identity of crime suspects. In the San Francisco laboratory, GC analyses of oil samples from the Russian tanker were performed and compared to thousands of samples from throughout the Middle East.

Relevant sections from the actual U.S. Customs (now U.S. Customs and Border Protection) Laboratory method (USCL Method 27-47) used to analyze the samples are shown below:[46]

## SECTIONS FROM THE U.S. CUSTOMS LABORATORY METHOD FOR COUNTY-OF-ORIGIN ANALYSIS

3. Reagents and Apparatus

3.1 Hewlett/Packard Model 5890 Series II Plus capillary gas chromatograph with a split/splitless injector, FID, liquid carbon dioxide subambient cooling oven attachment, autosampler control module, and electronic pressure control

3.2 Hewlett/Packard Model 7673A autosampler fitted with a 10 μL syringe

3.3 Hewlett/Packard Chem Station and software (PC)

3.4 Gas chromatograph operating conditions.

3.4.1 Split/splitless injector with a cup injector liner

3.4.2 J&W DB-1, 60-m, capillary column, part number 122-1063. This nonpolar column has a 1-μm film thickness and an internal diameter of 0.25 mm

3.4.3 Compressed helium carrier gas with a regulator for the capillary column

3.4.4 Compressed air with a regulator for the FID detector

3.4.5 Compressed hydrogen with a regulator for the FID detector

3.4.6 Compressed carbon dioxide with a liquid dip tube feeder to provide liquid carbon dioxide cooling to the gas chromatograph oven

3.4.7 Supelco High Capillary Gas Purifier

3.4.8 Supelco Thermogreen LB-2 septum

3.8 Reference samples:

3.8.1 Gasoline

3.8.2 Kerosene

3.8.3 Crude oil

3.8.4 Distillate fuel oil

3.8.5 Crude oil samples from oil fields of relevance to the instant analytical sample.

4. Sample Preparation

4.1 To prepare a sample of distillate petroleum for the gas chromatograph, 100 μL of distillate petroleum and 900 μL of cyclohexane are measured into a GC vial and capped.

4.2 Vortex the mixture for 15 s.

5. Experimental Procedure

5.1 The autosampler is loaded to run a set of samples such that each of the first two vials, the last vial, and one vial between each sample vial is a blank vial of cyclohexane. This provides significant evidence that the gas chromatograph is operating properly both before and after the analysis of any specific sample.

5.2 A 1.0 μL sample of the mixture is injected into the gas chromatograph via an autosampler using a 10-μL syringe.

5.3 Gas chromatograph run conditions.

5.3.1 The helium column carrier gas is under electronic pressure control with a column flow rate set at 1.5 ± 0.1 mL/min at 30 °C. The helium carrier gas is purified through an electrically heated Supelco High-Capacity Gas Purifier.

5.3.2 Split/splitless injector (cup injector liner packed with 10% OV-1 on Chromosorb-W) is used in the split mode at a split vent volume of $140 \pm 2.0$ mL/min.

5.3.3 The temperature of the injection port is isothermally maintained at 275 °C.

5.3.4 The oven temperature controller is set to hold the oven temperature for 5 min at 30 °C via liquid carbon dioxide subambient cooling of the oven and then to ramp the oven at 5 °C/min to 275 °C and hold it at that temperature for

a. 6 min for gasoline range material;

b. 40 min for mid-range distillate material;

c. 90 min for crude oil.

5.3.5 The temperature of the detector is isothermally maintained at 275 °C.

The analysis conditions described in the method illustrate many of the concepts discussed throughout this chapter. For example,

1. Proper controls and blanks (see Sections 3.8 and 5.1 in the USCL method) are used to verify system performance and to check for the presence of ghost peaks.

2. A split injection is used to avoid overloading the stationary phase with sample. This is important with such a thin stationary phase (1 µm) (see Section 5.3.2).

3. Temperature programming is used because oil sample components have a wide range of boiling points. The final temperature of 275 °C is held for a long time to elute low-volatility components to reduce the possibility of ghost peaks (see Section 5.3.4).

4. Helium is used as the carrier gas, which is quite common. Also, the common FID detector is used. This is an appropriate detector for oil samples because the components are hydrocarbon-based and therefore burn well in an FID to produce measurable signals (see Sections 3.4.3–3.4.5).

5. Kovats retention indices are used to identify key components (see Figure 2.30).

6. The use of a long (60 m), narrow-bore column (0.25 mm i.d.) with a thin film (1 µm) that is operated near its optimum flow rate (1.5 mL/min) maximizes the number of theoretical plates on the column to keep peaks as narrow as possible. This is important when analyzing complex mixtures such as oil that contain hundreds of closely related species that can overlap if columns with low plate counts are used (see Section 3.4.2).

The country of origin of an unknown sample is determined by comparing the chromatogram of the unknown to chromatograms obtained by analyzing thousands of samples from known sources and locations under the same conditions. In this analysis, the complete presence or absence of a peak in the chromatograms can help indicate the country of origin. However, the *ratio* of key peaks is also a critical indicator of whether or not an unknown sample matches one of the known samples.

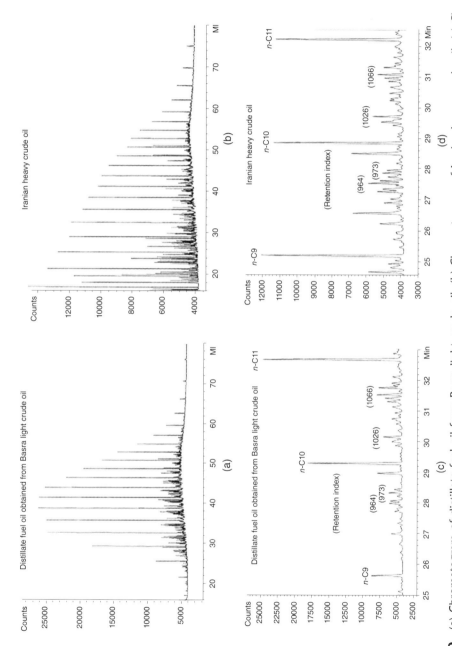

**FIGURE 2.30** (a) Chromatogram of distillate fuel oil from Basra light crude oil. (b) Chromatogram of Iranian heavy crude oil. (c) Chromatogram between nonane and undecane for distillate fuel oil from Basra light crude oil. (d) Chromatogram between nonane and undecane for Iranian heavy crude oil. These chromatograms illustrate that light crude oil from Basra is chemically different than heavy crude oil from Iran and that GC can be used to distinguish them. (*Source:* Reprinted with permission from U.S. Customs and Border Protection and Dr. Neal Byington.)

For example, consider the chromatograms in Figure 2.30. Comparing Figure 2.30 c and d, it is clear that Basra (Iraqi) light crude oil is chemically distinct from Iranian heavy crude oil in terms of the ratios of nonane:undecane, decane:undecane, and nonane:decane. Furthermore, the peaks whose Kovats retention indices have been labeled also offer distinct ratios.

The samples of oil suspected to be of Iraqi origin that were taken from the Russian tanker were analyzed by gas chromatography using the method described above. The chromatograms thus obtained were compared both visually and mathematically to the known samples, and a determination of the country of origin made.

In this case, 20% of the tanker's cargo was found to be Iraqi oil, in violation of the UN sanctions in place at that time.[47] Royal Dutch Shell maintained that the tanker was carrying only Iranian oil, but paid a $2 million fine and retained the cargo. In other incidences, the cargo and the tanker have been seized, sold on the open market, and the proceeds used to support U.N. relief programs. Given that a full tanker can carry oil worth tens of millions of dollars, having a shipment and tanker seized and sold represents a significant loss.

This application of gas chromatography points out the extreme importance of having a reliable technique for separating the components of complex mixtures. In this case, international law and political stability depended on it and the scientists at the U.S. Customs and Border Protection labs used gas chromatography to achieve it.

**EXAMPLE 2.14**

An oil sample of unknown origin is analyzed via the USCL protocol. The ratios of nonane to decane and decane to undecane are found to be 1.15 and 0.94, respectively. Is the sample more likely Iranian heavy crude oil or distillate fuel oil obtained from Basra light crude oil?

**Answer:** Iranian heavy crude because the ratios are closer to those in the chromatogram in the figure. In Basra oil, the ratio of nonane to decane and decane to undecane are both much less than 1.00. Of course, a more detailed comparison of all of the peaks present and their ratios would be necessary to increase the confidence of the conclusions drawn from the analysis. The USCL protocol calls for these more complete and complex analyses as part of the analysis process.

## 2.7. SUMMARY

Gas chromatography has existed for many decades because it is robust and reliable. As a result, gas chromatographs are one of the most common laboratory instruments. GC can be applied to the analysis of a wide variety of compounds to solve chemical problems, as demonstrated by the country-of-origin oil analysis discussed in this chapter. The retention and separation of analytes by gas chromatography is governed by intermolecular interactions between the analytes and the stationary phase. The result is that the two main instrumental factors controlling GC separations are the chemical nature of the stationary phase and the temperature at which the analysis is conducted. In practice, it is instrumentally easier to vary the temperature than it is to change the column, so temperature is usually the first variable that is adjusted when optimizing a separation. Temperature programming is used in the analysis of mixtures containing compounds of widely varying volatility. In addition to chemically different stationary phases (i.e., nonpolar, mid-polarity, and polar), different types of columns such as packed, support-coated, megabore, and capillary columns are used in gas chromatography, each with specific advantages for different applications. We have also seen that gas chromatographs generally consist of a carrier gas supply, an injection port, a temperature-controlled oven, and a detector. The injection process is a critical one in GC, as many analysis errors or poor performance characteristics can be traced to it. Internal standards are commonly used to improve precision in quantitative GC. The two most common detectors are flame ionization and thermal conductivity detectors, but several others such as nitrogen phosphorus detectors and electron capture detectors, which are more sensitive to or selective for specific types of analytes, are also routinely used. Advances in the field of gas chromatography include high-speed gas chromatography, microfabricated columns, and tandem GC. While new chromatographic and separation methods continue to be developed, it is almost certain that gas chromatography will remain one of the main instrumental methods within separation science.

## PROBLEMS

**2.1** Draw the structures of *N,N*-dimethylaniline and cumene (one polar and one not so). Which compound would you expect to be better retained on a Carbowax column? Explain.

**2.2** Draw the structures of benzylalcohol and a 50% cyanopropylmethyl–50% phenyl-methyl polysiloxane column. List the types of intermolecular interactions that benzylalcohol would experience on this column.

**2.3** Which column(s) (i.e., types of stationary phases) would most likely lead to a large separation factor between

(a) anisole and ethylbenzene?
(b) methylcyclohexane and toluene?

Explain your reasoning.

**2.4** Draw the structures of isopropanol and hexafluoroisopropanol (HFIPA).

(a) Which do you expect to be a stronger hydrogen bond donor, isopropanol or hexafluoroisopropanol? Explain your reasoning.
(b) Sketch a picture depicting the hydrogen bonding between hexafluoroisopropanol and both a cyanopropyl phase and a wax phase.
(c) On a DB-1701 phase (equivalent to an Rtx-1701) at 80 °C, $\log k$ for HFIPA was measured to be $-0.23$ and $\log k$ for IPA was $-0.73$. Calculate the separation factor for these two compounds on this phase at this temperature.

    (d) Will the separation factor in part (c) increase or decrease as the temperature increases. Explain your answer using intermolecular interactions as the basis for your reasoning.

**2.5**   (a) Draw the structures of phenol and *m*-cresol.

    (b) Predict where *m*-cresol elutes relative to phenol on all of the columns shown in Table 2.8. Explain your prediction.

    (c) Explain the increase in Kovats RI for phenol in going from DB-1 to DB-225 and wax phases.

**2.6**   (a) Based on their enthalpies of vaporization, predict the elution order of *n*-hexane, *n*-heptane, and methanol on a nonpolar column at 80 °C. The NIST website (http://webbook.nist.gov/chemistry/) may help with finding the required enthalpies of vaporization.

    (b) Actual retention factors on a polydimethylsiloxane (PDMS) column at 80 °C are as follows: methanol (0.0711), *n*-hexane (0.161), and *n*-heptane (0.343). Compare this order of elution to the one predicted by enthalpies of vaporization. Explain the discrepancy, basing your arguments on intermolecular interactions (hint: remember that the enthalpy of vaporization is a *bulk* solvent property).

    (c) Calculate the separation factors between methanol and *n*-hexane and between *n*-hexane and *n*-heptane.

    (d) What do you conclude about intermolecular interactions from all of the above calculations? Is the enthalpy of vaporization (or corresponding boiling points) always a good indication of relative retention? Why or why not? In what cases are boiling points going to provide good predictions and in which cases might they be misleading? Explain.

**2.7**   (a) In GC, does increasing the temperature lead to increased or decreased retention times?

    (b) Does increasing the temperature lead to increased or decreased separation factors?

**2.8**   Why might it be necessary to derivatize solutes for analysis by GC?

**2.9**   (a) Why can't enzymes and other large macromolecules be analyzed by GC?

    (b) Why can't ions (e.g., $Na^+$, $Mg^{2+}$, $Cl^-$, and acetate) be analyzed by GC?

**2.10**  What role does the carrier gas play in GC? How can carrier gas flow rates affect peak resolution?

**2.11**  It was asserted in the chapter that a plot of $\log k$ versus carbon number for the *n*-alkanes is linear. The table below has retention data for the *n*-alkanes collected on a 95%-methyl-5%-phenyl polysiloxane phase at 80.0 °C with a hold-up time of 0.275 min.

    (a) Calculate $\log k$ for each *n*-alkane.

    (b) Plot $t_r$ versus carbon number.

    (c) Plot $\log k$ versus carbon number and use linear regression analysis to determine the slope of the plot and subsequently the free energy of partitioning of a methylene group at 80 °C on this phase.

(d) Using the equation for the best straight line, determine $\log k$ and $t_r$ for undecane on this phase under these conditions.

| n-Alkane | $t_r$ (min) | $\log k$ |
|---|---|---|
| Octane | 0.705 | |
| Nonane | 1.169 | |
| Decane | 2.111 | |
| Undecane | | |
| Dodecane | 8.179 | |
| Tridecane | 16.703 | |
| Tetradecane | 34.200 | |

**2.12** A laboratory is going to be using GC for a routine analysis with hundreds of samples to analyze. They have developed two methods. One method is an isothermal method that requires 15 min to complete with an additional 1 min between analyses for syringe washes. A temperature programmed method also exists that elutes the compounds in only 12 min. This method, however, requires 6 min between analyses for the oven and column to cool down and to come to equilibrium at the initial temperature (during which time syringe washes could be performed).

(a) First, ignoring the time between analyses, how many samples could be analyzed by both methods in 1 week, assuming constant operation?

(b) Now, taking the time between analyses into consideration, recalculate the number of analyses that can be performed with the two different methods.

(c) In practice, which method should the laboratory adopt in order to maximize sample throughput?

(d) If the laboratory receives $150 per analysis, how much money would they lose each week by adopting the slower method?

**2.13** What is the difference between a destructive and nondestructive method of detection and which method(s) in Table 2.5 are nondestructive?

**2.14** Draw the structures of the following molecules and suggest a GC detector that would be highly sensitive to them (there may be more than one answer for some):

(a) Malathion (a pesticide)

(b) Tetradecane

(c) Mustard gas (a nerve agent)

(d) Phenyl propyl ether

(e) A sample containing chrysene, phenanthrene, and benzo[a]pyrene

(f) Fluoxetine and bupropion (antidepressants – what are their trade names?)

**2.15** What are the rare gases? What kind of detector and column (PLOT, SCOT, WCOT, or packed) would you use to analyze them?

**2.16** Consider the following average peak areas obtained via triplicate analysis for the purpose of preparing a calibration curve for methyl *tert*-butylether (MTBE) in gasoline. Each standard solution was prepared with 1% butyl ethyl as an internal standard.

| % MTBE in standard solution | Peak area MTBE | Peak area Butylether | Peak area ratio |
| --- | --- | --- | --- |
| 0.25 | 1025 | 4097 | |
| 0.50 | 1975 | 3955 | |
| 1.00 | 3997 | 4000 | |
| 1.25 | 5365 | 4280 | |
| 1.50 | 6557 | 4358 | |

(a) Considering only the MTBE peak areas, determine the MTBE content of an unknown sample if the MTBE peak area is found to be 5289 in the unknown. To do so, use the equation for the best straight line through the data determined using linear regression analysis.

(b) Using appropriate equations for the propagation of error when using calibration curves and assuming no uncertainty in the percent MTBE in each standard solution, determine the standard deviation in the value you determined in (a) assuming that the unknown was analyzed three times. (See Ref. 48 for propagation of error with linear regression results.)

(c) Calculate the ratio of MTBE to butylether peak areas, plot the data, and determine the equation for the best straight line through the data using linear regression analysis.

(d) Determine the percent MTBE in an unknown (i.e., peak area 5289) but also consider that the unknown contained 1% butylether internal standard, which yielded a peak area of 4235.

(e) Again using appropriate equations for the propagation of error when using calibration curves, assuming no uncertainty in the percent MTBE in each standard solution, determine the standard deviation in the value you determined in part (d). Compare the percent uncertainty in the MTBE content determined with and without the internal standard.

2.17 A column has an inlet pressure of 181.8 kPa and an outlet (ambient) pressure of 101.0 kPa. A flow rate of 5.20 mL/min is measured at the end of the column using a soap bubble meter. The column temperature is 100 °C and the ambient temperature (i.e., of the soap bubble meter) is 22 °C. Under these conditions, methane has a retention time of 0.43 min and hexylbenzene has a retention time of 3.73 min. Calculate the following:

(a) $F^{\circ}_{corr}$
(b) The corrected retention volume for methane and hexylbenzene
(c) The net retention volume for hexylbenzene.

Hint: You will need to look up the vapor pressure of water at 22 °C.

2.18 What is meant by "injection discrimination" and "inlet discrimination"? If discrimination occurs, does that introduce a systematic bias or random uncertainty into the amount of solute injected compared to its actual concentration in the sample?

2.19 Describe the difference in the injection port valve settings in split and splitless injections. Which is used for concentrated samples?

**2.20** What is "carryover" and how can it be minimized in GC?

**2.21** Internal standards improve the precision of GC analyses. What is it about the injection process that makes it desirable to use internal standards in GC analyses?

**2.22** What does SPME stand for and why is it used? Describe how it works.

**2.23** To a first approximation, the response of flame ionization detectors is proportional to the number of C—H bonds in the molecule. Suppose that in the analysis of a sample, the following raw peak areas were obtained:

| Analyte | Peak area (arbitrary units) | Raw area percent | Actual percent |
|---|---|---|---|
| *n*-Hexane | 67,000 | | |
| *n*-Propyl benzene | 70,000 | | |
| Naphthalene | 54,000 | | |
| *t*-Butylnaphthalene | 67,000 | | |

(a) *Without* taking detector sensitivity into consideration, determine the percent area for each analyte in the sample (i.e., analyte area/total area).

(b) Now, keeping in mind that FID signal is roughly proportional to the number of C—H bonds in the molecule, recalculate the actual percentage of each component in the sample. (Use *n*-hexane as a reference to calculate the signal per C—H bond and recalculate peak areas using that reference but also considering the ratio of C—H bonds in each analyte.)

(c) Comparing the percentages based on raw peak area versus those that consider detector sensitivity, what can you conclude about the accuracy of using unadjusted peak areas for determining the percent composition of a sample?

**2.24** Based on the following chromatogram, answer the following questions. Assume the chromatogram was collected under isothermal conditions using pentane as a solvent. The hold-up time was independently measured to be 18.0 s using methane in a separate injection.

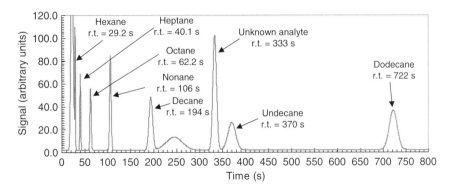

(a) What is the Kovats retention index of the unknown analyte?

(b) If the unknown analyte is identified to be benzaldehyde and the column used in this example was a cyanopropylphenyl phase, would you expect the Kovats retention index to increase or decrease if the same sample were injected on a diphenyldimethylpolysiloxane phase?

(c) From what does the first, large, early-eluting, and off-scale peak arise? Why is it so much larger than the other peaks?

(d) If the peak that appears around 240 s appears to "move around" on the chromatogram (i.e., shows up at different retention times) when a sequence of repeat injections is done, what would you call this type of peak and why do they arise? Why is it significantly broader than the other peaks around it? How should the analysis be changed to prevent it from appearing to move around?

(e) Why doesn't the retention time increase linearly with the number of carbons in the homologous series of *n*-alkanes? What parameter does increase linearly with carbon number?

(f) Hexane overlaps with the first peak. What parameter(s) would you change to achieve a better separation?

(g) If you were not interested in calculating Kovats retention indices and simply wanted to separate and quantify the compounds in this mixture (including the *n*-alkanes), would you continue using an isothermal method. List two *potential* benefits that using a temperature programmed analysis might have. What potential drawbacks might it have?

2.25 In the oil analysis described at the end of the chapter,

(a) what kind of stationary phase was used?

(b) is this a polar or nonpolar column?

(c) were the samples that were being analyzed generally comprised of polar or nonpolar compounds?

(d) what were the column dimensions and stationary phase thickness?

(e) what kind of detector was used? Why was this an appropriate choice?

(f) what kind of injection method was used (i.e., split, splitless)?

(g) was the analysis isothermal or temperature programmed?

2.26 Explain how the inclusion of step 5.1 in the oil analysis experimental procedure provides "significant evidence that the gas chromatograph is operating properly both before and after the analysis of a specific sample" as the analysts assert. If you were the person analyzing the samples, what would you expect to see as a result of step 5.1?

2.27 Polyhalogenated biphenyls (PCBs) are common environmental pollutants but are often found at trace levels. Find a journal article or website that *details* the analysis of PCBs in aqueous samples by gas chromatography. Note the following: sample preparation and any preconcentration steps (how and why are these conducted), choice of GC conditions (e.g., nature of stationary phase, dimensions of column, oven temperature, temperature programming), and choice of detector. Explain the need for these steps or conditions and propose reasons for the specific decisions made by the analysts. In other words, put yourself in their position and defend the method. If better alternatives exist, say so.

## REFERENCES

1. Davankov, V.A.; Dominguez, J.A.G.; Diez-Masa, J.C. *Pure Appl. Chem.* 2001, *73*, 969–992.

2. Davankov, V.A. *Chromatographia* 2003, *57*, s195–s198.

3. Davankov, V.A.; Onuchak, L.A.; Kudryashov, S.Y.; Arutyunov, Y.A. *Chromatographia* 1999, *49*, 449–453.

4. Davankov, V.A. *Chromatographia* 1998, *48*, 71–73.

5. Parcher, J.F. *Chromatographia* 1998, *47*, 570–574.

6. Hinshaw, J.V. *LC/GC North Am.* 2002, *20*, 948–952.

7. Grob, R.L.; Barry, E.F. *Modern Practice of Gas Chromatography*, 4$^{th}$ Ed.; John Wiley & Sons, Inc.: Hoboken, 2004.

8. Jennings, W.; Mittlefehldt, E.; Stremple, P. *Analytical Gas Chromatography*, 2$^{nd}$ Ed.; Academic Press: New York, 1997.

9. Grob, K. *Split and Splitless Injection for Quantitative Gas Chromatography: Concepts, Processes, Practical Guidelines, Sources of Error*, 4$^{th}$ Ed.; Wiley-VCH: New York, 2001.

10. See commercial chromatographic supplier catalogs for descriptions, common usage, and advantages/disadvantages of different injection port liners.

11. Klee, M.S. *GC Inlets – An Introduction*; Hewlett-Packard Co., 1991.

12. Robard, K.; Haddad, P.R.; Jackson, P.E. *Principle and Practice of Modern Chromatographic Methods*; Academic Press: New York, 1994.

13. Sandra, P. (Ed.) *Sample Introduction in Capillary Gas Chromatography*, Vol. 1; Heutig: Heidelberg, 1985.

14. Arthur, C.L.; Pawliszyn, J. *Anal. Chem.* 1990, *62*, 2145–2148.

15. Pawliszyn, J., *Solid Phase Microextraction: Theory and Practice*; Wiley-VCH: New York, 1997.

16. Buffington, R.; Wilson, M.K. *Detectors for Gas Chromatography – A Practical Primer*; Hewlett-Packard, 1991.

17. Dietz, W.A. *J. Gas Chromatogr.* 1967, *5*, 68–71.

18. Scanlon, J.T.; Willis, D.E. *J. Chromatogr. Sci.* 1985, *23*, 333–340.

19. Hinshaw, J.V. *LC/GC Europe* 2006, *19*(6), 344–351.

20. McNair, H.M.; Miller, J.M. *Basic Gas Chromatography*; John Wiley & Sons, Inc.: New York, 1998.

21. Patterson, P.L. *J. Chromatogr. Sci.* 1986, *24*, 41–52.

22. Martin, A. J. P. *Biochem. Soc. Symp.* 1949, *3*, 4–20.

23. Vailaya, A.; Horváth, Cs. *J. Phys. Chem. B* 1998, *102*, 701–718.

24. Kovats, E.S. *Helv. Chim. Acta* 1958, *41*, 1915–1932.

25. McReynolds, W.O. *J. Chromatogr. Sci.* 1970, *8*, 685–691.

26. McReynolds, W.O. *Gas Chromatographic Retention Data*, 5$^{th}$ Ed.; Preston: Niles, 1987.

27. Rohrschneider, L. *J. Chromatogr.* 1966, *22*, 6–22.

28. Vitha, M.F.; Carr, P.W.; Mabbott, G.A. *J. Chem. Educ.* 2005, *82*, 901–902.

29. Miller, J.M., *Chromatography: Concepts and Contrasts*, 2$^{nd}$ Ed.; John Wiley & Sons, Inc.: Hoboken, 2005.

30. Poole, C.F., Zlatkis, A. *Anal. Chem.* 1980, *52*, 1002A–1016A.

31. Perry, J.A.; Feit, C.A. "Derivatization Techniques in Gas–Liquid Chromatography" in *GLC and HPLC Determination of Therapeutic Agents, Part I*, Tsuji, K. and Morozoawich, W. (Eds.); Dekker: New York, 1978.

32. Drozd, J. *Chemical Derivatization in Gas Chromatography*; Elsevier: Amsterdam, 1981.

33. Blau, K.; Halket, J.M. (Eds.) *Handbook of Derivatives for Chromatography*, 2nd Ed.; Hoboken: John Wiley & Sons, Inc., 1993.

34. Sacks, R.D.; Dimandja, J-M.D., Jr.; Patterson, D.G., Jr., *High Speed Gas Chromatography*; John Wiley & Sons, Inc.: Hoboken, 2001.

35. van Es, A. *High-Speed Narrow Bore Capillary Gas Chromatography*; Heutig Buch Verlag: Heidelberg, 1992.

36. Grall, A.; Leonard, C.; Sacks, R. *Anal. Chem.* 2000, *72*, 591–598.

37. Sacks, R.; Smith, H.; Nowak, M. *Anal. Chem.* 1998, *70*, 29A–37A.

38. Smith, H.; Sacks, R. *Anal. Chem.* 1998, *70*, 4960–4966.

39. Purnell, J.; Wattan, M. *Anal. Chem.* 1991, *63*, 1261–1264.

40. Akard, M.; Sacks, R. *Anal. Chem.* 1994, *66*, 3036–3041.

41. Jones, J.; Purnell, J. *Anal. Chem.* 1990, *62*, 2300–2306.

42. Sacks, R.; Coutant, C.; Grall, A. *Anal. Chem.* 2000, *72*, 524A–533A.

43. Lambertus, G.; Elstro, A.; Sensenig, K.; Potkay, J.; Agah, M.; Scheuering, S.; Wise, K.; Dorman, K.; Sacks, R. *Anal. Chem.* 2004, *76*, 2629–2637.

44. Steinecker, W.H.; Rowe, M.P.; Zellers, E.T. *Anal. Chem.* 2007, *79*, 4977–4986.

45. Associated Press, 2000. http://www.newsmax.com/articles/?a=2000/4/7/140018.

46. Byington, N.D. and Fluty, L.D. U.S. Customs Laboratory Methods, USCL Method 27-47. Guidelines for Country-of-Origin Determinations of Distillate Petroleum Products from Iraq.

47. Myers, S.L.; Fining Shell, UN Concludes That Tanker Carried Iraq Oil, *New York Times,* April *26*, 2000.

48. Harvey, D. *Modern Analytical Chemistry*; McGraw-Hill Science: Dubuque, IA, 1999.

**FURTHER READING**

1. Buffington, R.; Wilson, M.K. *Detectors for Gas Chromatography – A Practical Primer*; Hewlett-Packard, 1991.

2. Grob, K. *Split and Splitless Injection for Quantitative Gas Chromatography: Concepts, Processes, Practical Guidelines, Sources of Error*, 4th Ed.; Wiley-VCH: New York, 2001.

3. Jennings, W.; Mittlefehldt, E.; Stremple, P. *Analytical Gas Chromatography*, 2nd Ed.; Academic Press: New York, 1997.

4. McNair, H.M.; Miller, J.M. *Basic Gas Chromatography*; John Wiley & Sons, Inc.: New York, 1998.

5. Miller, J.M., *Chromatography. Concepts and Contrasts*, 2nd Ed.; John Wiley & Sons, Inc.: Hoboken, 2005.

6. Poole, C.F.; Poole, S.K. *Chromatography Today*; Elsevier: Amsterdam, 1991.

7. Robard, K.; Haddad, P.R.; Jackson, P.E. *Principle and Practice of Modern Chromatographic Methods*; Academic Press: New York, 1994.

8. Sandra, P. (Ed.) *Sample Introduction in Capillary Gas Chromatography*, Vol. 1; Heutig: Heidelberg, 1985.

# LIQUID CHROMATOGRAPHY

**3**

The importance of liquid chromatography (LC) as an analytical technique and its impact on our lives cannot be overstated. It is used to separate, identify, and quantify amino acids, environmental toxins, active ingredients in medication, drugs of abuse, surfactants, carbohydrates, and many other compounds of extreme importance. These types of samples often cannot be analyzed by gas chromatography because the compounds are not sufficiently volatile or are thermally labile. In fact, it has been estimated that only about 20% of organic compounds can be analyzed by gas chromatography. Liquid chromatography is therefore a very important separation technique. We begin this chapter with two examples of how liquid chromatography is used in practical applications, specifically to screen food for contaminants and to test athletes for performance-enhancing drugs. Then, we discuss the history, theory, and instrumentation upon which LC is based. We end the chapter by describing the application of liquid chromatography to monitor pharmaceutical compounds in groundwater throughout the United States.

## 3.1. EXAMPLES OF LIQUID CHROMATOGRAPHY ANALYSES

The quality and safety of our daily food supply is highly dependent on liquid chromatography. Many cases of food contamination – either deliberate or naturally occurring – have made the news over the past several years. Common examples include the deliberate addition of melamine in milk and infant formula in China, and the addition of carcinogenic Sudan dyes to chili powder that was then used in a number of other food products in the United Kingdom.

Melamine, a small molecule that has a high percentage of nitrogen atoms in it, was added to milk and infant formula because tests for protein content actually test for nitrogen, not the proteins themselves. Spiking milk and baby formula with melamine enhances the apparent protein content, helping producers prevent their milk from being rejected from regulators, or helping command higher prices. The melamine made hundreds of thousands of people sick, tens of thousands were hospitalized, and several infants died.

*Chromatography: Principles and Instrumentation*, First Edition. Mark F. Vitha.
© 2017 John Wiley & Sons, Inc. Published 2017 by John Wiley & Sons, Inc.

Sudan dyes are added to chili pepper powder to enhance the color of the spice. Because chili pepper is evaluated partly on appearance, enhancing the color allows for higher prices to be charged. Unfortunately, the dyes are carcinogenic and are therefore banned for food use. Yet, chili pepper tainted with Sudan dyes turned up in relishes, chutneys, and seasonings in the United Kingdom and other European Union countries.[1,2] The origin was chili pepper that had been imported from India into Europe. No direct deaths or injuries were reported in this case, yet it is another example of a massive, deliberate contamination of food.

The United States Food and Drug Administration (US FDA) monitors hundreds of substances in food products. Currently, the state of the art is to use ultra-high performance liquid chromatography (UHPLC) to separate components in food samples and analyze the compounds as they elute with high-resolution mass spectrometry (HRMS). In addition, techniques such as hydrophilic interaction liquid chromatography (HILIC) are being used to develop methods for the monitoring of polar contaminants such as melamine, mentioned earlier.[3,4] Every day, FDA laboratories are using hundreds of liquid chromatography systems to screen thousands of samples, looking for potentially harmful substances in an effort to keep them out of the food supply. If these screening methods detect possible contamination or harmful substances, additional, more specific and stringent confirmatory tests – many of which also involve chromatography – are used to verify the presence of suspected contaminants and to quantify them if present.

Another important application of liquid chromatography is in drug testing for athletes (e.g., the Olympics, NCAA participation, Tour de France). The World Anti-Doping Agency has specified over 400 substances as being prohibited (see www.wada-ama.org). In order to enforce these bans, it is important to develop analytical methods capable of detecting and quantifying these compounds as well as hundreds of related species such as their biological precursors and metabolites. Throughout all sports, thousands of athletes, at all levels, are screened each year for a variety of banned substances.

Some of the substances that receive considerable attention in the news include steroids such as testosterone and epitestosterone, and precursors such as androstenedione ("andro"). The steroids help build muscle and are therefore attractive doping agents for certain athletes. Another performance-enhancing compound of interest is erythropoietin (EPO), a hormone that increases the production of red blood cells and hence the oxygen-carrying ability of blood. It is therefore used by some endurance athletes and has been in focus as a popular doping agent in the world of professional cycling, including the Tour de France, for many years.

One of the challenges of enforcing bans on these and other substances is that many of them are produced naturally in the body. Thus, it is not enough to detect their presence, but to determine if artificially elevated levels are present. Another challenge for drug testing is that, ideally, the analyses should be as rapid as possible. This way, penalties for those who are found to be doping can be applied in a timely manner.

Just as we saw for food screening, liquid chromatography coupled with mass spectrometry (LC-MS) is an excellent combination for the task of rapidly screening large numbers of samples for a wide variety of compounds.[5] For example, Jeong et al. reported a method for analyzing urine samples based on ultrafast liquid chromatography with triple quadrupole mass spectrometry.[6] After completing the sample preparation, the analysis could screen for 210 prohibited substances in 10 min. Dominguez-Romero et al. also reported the rapid screening of 200 sport drugs in urine samples using fast liquid chromatography with a time-of-flight mass spectrometer as a detector.[7] Target compounds in this study included testosterone and epitestosterone.

When screening methods such as those mentioned above detect possible cases of doping, more detailed confirmatory tests are required. In many cases, GC-MS is used for this purpose. Also, because many of the dopants are now more complex biological molecules such as proteins and hormones, techniques such as electrophoresis are increasingly being used. For example, confirming the presence of exogenous EPO (also known as recombinant erythropoietin or rEPO) and distinguishing it from naturally occurring EPO is performed by one of two electrophoretic methods: SDS-PAGE or SAR-PAGE (sodium dodecyl sulfate–, or sarcosyl–polyacrylamide gel electrophoresis).[8] In this way, athletic drug screening – just like food screening – highlights the primary role played by liquid chromatography as a method for rapidly, accurately, and sensitively detecting a wide range of compounds of interest. It also points to the need for a whole host of analytical techniques, each with particular strengths, to confirm results. In both food analysis and sports, careers, lives, and significant amounts of money are at stake, so it is critical that accusations of food contamination or drug doping are based on reliable analytical results.

## 3.2. SCOPE OF LIQUID CHROMATOGRAPHY

In liquid chromatography, a liquid mobile phase is pumped through a column packed with highly porous particles. This mobile phase transports analyte molecules through the column. The surface of the particles (primarily the surface *inside* the pores) or molecules chemically bonded to the surface (the stationary phase) interact with the analyte molecules and cause them to be retained in the column. The total retention time of any solute is therefore dictated by its affinity for the mobile phase *relative* to its affinity for the stationary phase. Molecules that are highly attracted to the mobile phase and not attracted to the stationary phase are only slightly retained, while molecules that have strong interactions with the stationary phase and little affinity for the mobile phase are highly retained. Complex mixtures of molecules can thus be separated because different compounds have different blends of mobile phase and stationary phase affinities.

Because both the mobile and stationary phases interact with the analytes, controlling retention in LC is different than controlling retention in GC, where the mobile phase is essentially noninteracting. Furthermore, LC is different from GC because the molecules are transported in a liquid in LC, not by a gaseous mobile phase as in GC. Thus, analytes do not need to be volatile in order to be analyzed. This means that LC can be used to analyze a much wider range of compounds compared to GC. These include large molecules such as polymers, peptides, proteins, carbohydrates, and steroids, as well as smaller but still nonvolatile analytes such as amino acids, carboxylic acids, organic bases, and metal ions.

**FIGURE 3.1**  M.S. Tswett in Keil, Germany, in 1905. (*Source:* Excerpted and reprinted with permission from *LC/GC North Am.*, 2006 (24) 680. *LC/GC North America* is a copyrighted publication of Advanstar Communications Inc. All rights reserved.)

## 3.3.  HISTORY OF LC

The first report of LC separations came from Mikhail Tswett, a Russian botanist, in 1903 (Figure 3.1).[9] He used large diameter glass columns filled with solid powders to separate pigments extracted from plants (carotenoids and chlorophyll). To do so, he continuously added solvent to the column after introducing the pigment extract at the top of a vertical column. The pigments adsorbed differentially onto the solid and separated as they migrated down the column. As the pigments separated, colored rings appeared within the column, each ring being a different pigment extracted from the leaves. It is for this reason that the term *chromatography* (chroma = color and graphy = to write; i.e., writing in color) is used. Coincidentally, the name Tswett means "color" in Russian, so the term chromatography could also be interpreted as "Tswett's writing."[9]

As often happens, Tswett's work did not gain immediate acceptance. It was not until the 1930s that the technique was resurrected. At that time, however, it suffered from three main drawbacks: broad solute bands that often overlapped, the slow and tedious nature of the process, and the difficulty in detecting colorless species. This situation lasted for nearly 30 years.[9] Then, between 1964 and 1966, Csaba Horvath, working at Yale University, fundamentally changed the way liquid chromatography was done. Theoretical work, especially that of Giddings,[10,11] suggested that the use of small particles of micron dimensions packed tightly in narrow columns would produce much narrower solute bands than large diameter columns packed with large particles. Small particles, however, impede solvent flow, which makes using gravity-based flow essentially impossible due to exorbitantly long analysis times. Horvath overcame the problem by forcing the mobile phase through the packed column using high-pressure liquid pumps rather than relying on gravity. In addition, at the end of the column, he placed a flow-through UV detector to allow detection of colorless analytes.[12]

This instrument simultaneously overcame the broad bands, slow analysis times, and many of the detection difficulties. Because of the pressure pumps employed, the technique was sometimes referred to as high-*pressure* liquid chromatography (HPLC), and between 1970 and 1972, several commercial instruments became available. It is more common now to associate the acronym HPLC with high-*performance* liquid chromatography, emphasizing the enhanced resolution that can be achieved with modern chromatography systems.

### 3.3.1. Modern Packing Materials

While Tswett used alumina and calcium carbonate as his packing material, silica ($SiO_2$) particles are today's workhorse for liquid chromatography. Silica consists of silicon atoms bridged by oxygen atoms terminating with some SiOH groups at the surface as depicted in Figure 3.2.[13] The silica particles are spherical and porous. The diameter of the particles, the average pore size, and the pore size distribution significantly influence separations. An electron micrograph of typical silica particles is shown in Figure 3.3, and a schematic of the porous nature of the particle is shown in Figure 3.4.

Particles with a diameter of 5 μm are commonly used, but as discussed in Chapter 1 and in this section, the trend is toward using smaller particles because smaller particles provide narrower peaks (i.e., smaller plate heights). Common pore sizes range from 60 to 300 Å. The silica particles are between 45% and 55% porous, meaning that the vast majority of the particles' surface area is *inside* the pores, not on the exterior surface of the particle (imagine all the surface area on the inside of a Styrofoam ball compared to its external surface area). Because analytes are retained on the surface of the pores, *the pores are responsible for nearly all solute retention*. For typical silica particles used in routine analyses, 1 g of packing material has 300 $m^2$ of surface area, a remarkable amount of area for a single gram of material.

Much of the success of modern liquid chromatography is due to the work of Kirkland, Snyder, and Unger.[14,15] Their production of surface-modified silica particles with the specifications required for successful chromatography and their fundamental studies of

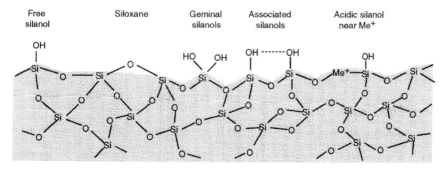

**FIGURE 3.2** Structure of silica. (*Source: Practical High-Performance Liquid Chromatography*, V. Meyer 3rd edition, p.105, 1998. Copyright John Wiley and Sons, Inc. Reproduced with permission.)

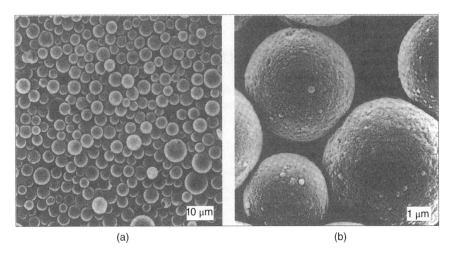

(a)                                                        (b)

**FIGURE 3.3** Scanning electron micrograph of spherical porous silica particles (Spherisorb octadecyl-silane chemically modified silica). Magnification 700× (a) and 7000× (b). Mean particle diameter 5 μm. (*Source: Practical High-Performance Liquid Chromatography*, V. Meyer 3$^{rd}$ edition, p.105, 1998. Copyright John Wiley and Sons, Inc. Reproduced with permission.)

**FIGURE 3.4** Depiction of the porous nature of typical HPLC silica particles (cross-sectional view). Gray-shaded regions represent the solid silica material, in this case with a stationary phase (short black curves) bonded to the surface. Open white spaces represent the pores.

solute retention led to a more complete understanding of HPLC and significantly helped to establish it as a major analytical technique.

It is important to get a mental image of the dimensions involved in HPLC packing materials. A typical particle is 5 μm, or $5 \times 10^{-6}$ m, in diameter. A typical pore diameter is 80–120 Å or $0.8$–$1.2 \times 10^{-8}$ m; thus, the pores are 400–600 times smaller than the particle itself. Given the spherical nature of the particles, this allows for thousands of pores to be formed in the particle. These pores provide the high surface area *inside* the particles that results in solute retention.

Because solutes are retained by the surface inside the pores, there must be a mechanism that brings the solutes into the pores and into contact with the surface. As we discussed in Chapter 1, solutes are brought to the exterior of the particles by the moving mobile phase. However, the solutes cannot be pushed into the pores by the flowing mobile phase because the pores are so small that they restrict the flow of the mobile phase through them. Thus, the mobile phase flows around the particles rather than through them. The pores themselves are filled with mobile phase that is stagnant. So solutes must diffuse from the flowing mobile phase into the stagnant mobile phase inside the pores. The solutes must subsequently diffuse through the stagnant mobile phase inside the pores to be retained by the stationary phase bonded to the pore walls. They must also diffuse through the stagnant mobile phase that fills the interconnected pores in order to exit the particle and thereby get back into the moving mobile phase to be transported further down the column (see Figure 1.21 on page 42). The question, then, is: "Can solutes fit into pores that are 120 Å in diameter?" To answer this, consider the fact that the average carbon–carbon bond length is 1.5 Å or $1.5 \times 10^{-10}$ m and that the hard-sphere diameter of benzene is approximately $4.5 \times 10^{-10}$ m.[16] Thus, the pore opening is approximately 25 times wider than common small organic molecules, so it is certainly possible for solutes to fit into the pores.

To put this in more familiar terms, imagine that the solute molecule is pea-sized (approximately 1.5 cm in diameter). The pores would then be approximately 37.5 cm (15 in.) in diameter, and the particle itself would be 17,000 cm (560 ft) in diameter. Thus, if the solute were a pea, the particle would be about a tenth of a mile in diameter. The particle now can be viewed as a giant sphere covering several city blocks pockmarked with thousands of inlets each about a foot in diameter, into which the pea-sized solute can diffuse. If you can visualize this, you have an approximate feeling for the scale of chromatography and the process that ultimately leads to solute retention.

It must be noted that while silica is the dominant packing material in use, it suffers from rapid dissolution in aqueous media above pH 7–8. Thus, if mobile phases with high pH values are used, the silica particles begin to dissolve, which ruins the column. It is therefore important to test the pH of mobile phases before use. Some recent advances in column technology and surface modifications have achieved greater stability above pH 8, but for most columns, high pH values still represent a significant limitation. A summary of the range of characteristics of silica particles is presented in Table 3.1.

While spherical silica particles are used for the vast majority of chromatographic separations, porous alumina, titania, and zirconia particles are also used in some applications. In addition, nonporous spherical particles made from organic polymers are also commercially available.

**TABLE 3.1   Characteristics of Common Silica Particles for Liquid Chromatography**

| Characteristics of porous silica particles | |
|---|---|
| Diameter | 1–300 μm (1–5 μm are common) |
| Pore size | 60–300 Å |
| Porosity | 45–55% |
| Surface area | 100–300 m²/g |
| pH stability range | 2–8 |

## 3.4.   MODES OF LIQUID CHROMATOGRAPHY

There are many modes in which liquid chromatography can be performed. Each fills a specific niche in terms of the types of solutes for which they are most suited. The most important modes are

1. normal phase liquid chromatography (NPLC);
2. reversed-phase liquid chromatography (RPLC);
3. ion-exchange chromatography (IEX);
4. hydrophilic interaction chromatography (HILIC);
5. size exclusion chromatography (SEC) [also called gel permeation chromatography (GPC)]; and
6. affinity chromatography.

The key characteristics of these modes are summarized in Table 3.2. They are discussed in detail in the following sections.

### 3.4.1.   Normal Phase Liquid Chromatography (NPLC)

When Tswett was using powders such as alumina and calcium carbonate as the stationary phase, his mobile phases consisted of organic solvents such as petroleum ether, ethanol, or acetone. In these cases, the inorganic stationary phase is more polar than the organic-based mobile phase. Because this was historically the first mode of chromatography, it has become known as normal phase liquid chromatography (NPLC). While it was the first mode, it is no longer the most common mode, having been surpassed for many reasons by reversed-phase LC, which is described in later sections.

The main mode of retention in NPLC is adsorption of solutes to a solid surface as depicted in Figure 3.5 (see also the cartoon depicting adsorption and absorption in Figure 2.26 on page 122 in Chapter 2). The extent of solute separation and adsorption depends on the balance of two sets of intermolecular interactions: attractive forces between the solutes and the surface compared to the attractive forces between the solute and the mobile phase. The surfaces in NPLC are polar, with silica, alumina, and titania being common. The mobile phases are relatively nonpolar – usually hexane modified with slightly polar solvents such as ethanol and acetone. Given the polar nature of the stationary phase and the less polar nature of the mobile phase, polar solutes tend to favor adsorption to the

**TABLE 3.2   Characteristics of Major Modes of Liquid Chromatography**

| Mode | Stationary phase | Mobile phase | Common analytes |
| --- | --- | --- | --- |
| Normal phase liquid chromatography (NPLC) | Polar (e.g., bare silica, alumina, titania) | Nonpolar (e.g., hexane modified with more polar solvents) | Neutral organic compounds of moderate size |
| Reversed-phase liquid chromatography (RPLC) | Nonpolar (e.g., silica modified with alkyl chains) | Polar (e.g., water modified with less polar solvents) | Neutral organic compounds of moderate size |
| Ion-exchange chromatography (IEX) | Charged polymeric phases (sulfonates, carboxylic acids, tertiary amines) | Aqueous buffers | Inorganic and organic ions |
| Hydrophilic interaction chromatography (HILIC) | Polar (e.g., bare silica, cyano phases, amino phases) | Nonpolar with a high percentage (>60%) of organic and very little water (2.5–40%) | Small, polar, and charged compounds like peptides that typically elute early in RPLC |
| Size exclusion chromatography (SEC) also called gel permeation chromatography (GPC) | Porous polymeric stationary phases | Aqueous with organic solvents | High molecular weight species such as polymers, proteins, and large biomolecules |
| Affinity chromatography | Immobilized biochemicals (antibodies, antigens, enzymes) | Aqueous with organic solvents | Specific enzymes, biomolecules, or targeted antigens |

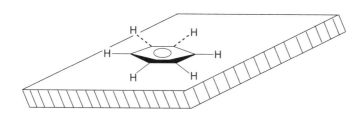

**FIGURE 3.5**   Depiction of the adsorption of benzene onto a surface.

surface whereas nonpolar solutes tend to favor remaining in the mobile phase. It is important to recall from Chapter 1 that retention is ultimately a thermodynamic process related to the adsorption coefficient, $K$. In NPLC, because the stationary phase is polar and the mobile phase is relatively nonpolar, the equilibrium $A_{mob} \Longleftrightarrow A_{stat}$ (where "$A$," "mob," and "stat" stand for analyte, mobile phase, and stationary phase, respectively) lies more toward the product side for polar solutes than for nonpolar solutes. This means that the equilibrium constant, $K$, is greater for polar solutes. Because the retention factor, $k$, is proportional to $K$ (recall $k = K/\beta$, where $\beta$ is the phase ratio) and is ultimately a measure of solute retention, it is clear that increasing $K$ by increasing the affinity of the solute for the stationary phase relative to its affinity for the mobile phase results in greater retention. Thus, in general, nonpolar solutes elute earlier than polar solutes in NPLC. Furthermore, because retention relies on solute/surface interactions and the specific geometry of those interactions, NPLC is sensitive to the relative positions of functional groups on a molecule. Therefore, NPLC is often used as an alternative to the reversed-phase mode discussed below for the separation of geometric isomers and large polyaromatic hydrocarbons (PAHs).

While solute/surface interactions are important in governing retention, it is important to remember that the column is also filled with mobile phase and that the mobile phase molecules compete with solutes for adsorption sites on the stationary phase surface. This competition reduces the overall retention of solutes. For example, a polar solvent such as methanol competes well with solute molecules for surface adsorption sites. Thus, a nonpolar solute such as anthracene elutes sooner in methanol than in a solvent such as hexane, even though its solubility is considerably lower in methanol than it is in hexane.

### 3.4.2. Reversed-Phase Liquid Chromatography (RPLC)

In contrast to the polar stationary phases and nonpolar mobile phases originally used by Tswett and which characterize NPLC, Horvath reversed the polarities of the mobile and stationary phases by using particles modified with nonpolar alkane-like molecules. In combination with this, he used polar mobile phases that were comprised primarily of water modified to some extent by polar organic solvents such as acetonitrile (ACN), methanol (MeOH), and tetrahydrofuran (THF). Because these nonpolar stationary phases coupled with polar mobile phases are the reverse of traditional chromatography, such systems are referred to as reversed-phase liquid chromatography (RPLC).

While chromatography has its roots in NPLC, RPLC has largely taken over, with the vast majority of LC separations being performed in the reversed-phase mode. There are, however, significant applications for NPLC, so both modes must be understood. Furthermore, while they are related, they operate on fundamentally different physical processes. Specifically, retention in NPLC is caused by *ad*sorption of the solutes *onto* the stationary phase surface while retention in RPLC is caused by solutes partitioning *into* a liquid-like stationary phase coated on the support particle surface (i.e., *ab*sorption).

In the most common mode of RPLC, the surface of the silica particles (including the extensive surface inside the pores) is modified with C-18 chains through chemistry shown in Figure 3.6. These phases are known as octadecylsilane, or ODS, phases. The stationary phase therefore acts somewhat like a liquid alkane into which solutes can diffuse. Because the solute is embedded in the stationary phase, the mechanism of retention is solute *partitioning* between the mobile and stationary phases (see Figure 3.7) as opposed to the adsorption process in NPLC. The mechanisms are similar, however, because the extent to which a solute is retained in both RPLC and NPLC depends on the strength of the solute's intermolecular interactions with the mobile phase compared to the strength of its interactions with the stationary phase. For example, consider the retention of phenol and benzene. If the stationary phase is assumed to be alkane-like, then the stationary phase can interact only through dispersion and dipole–induced dipole forces arising from a polar solute inducing a dipole in the stationary phase. In the case of phenol and benzene, predicting the exact strength of interactions with the stationary phase is difficult, but because phenol is larger, it is reasonable to think it may have stronger dispersion

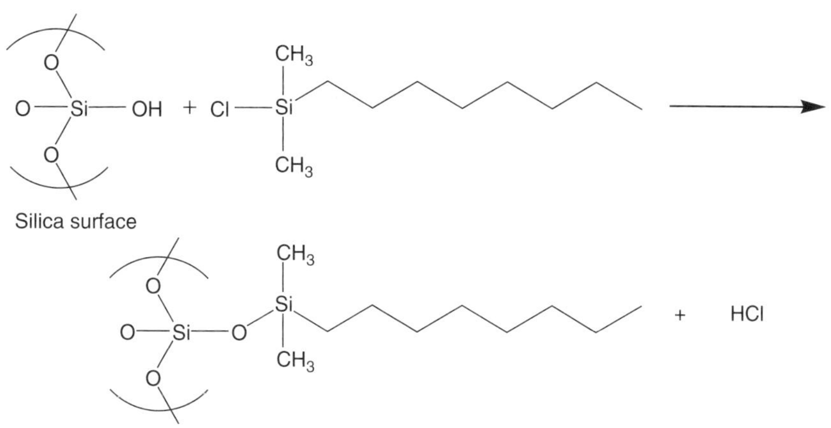

**FIGURE 3.6** Depiction of the chemistry for modifying a silica stationary phase with an octylsilane (C-8) linkage. Many other surface modification techniques also exist.

interactions. Furthermore, it is also polar, so it can induce a dipole in the stationary phase chains. Benzene, conversely, is slightly smaller but will still interact through dispersion interactions comparable in strength to those of phenol. But because it lacks a permanent dipole it cannot induce a dipole in the stationary phase. On the basis of this analysis, one might predict that phenol is slightly more retained due to the additional mode of interaction, or at the very least, retention for the two compounds is quite similar. *However, the intermolecular interactions between these solutes and the mobile phase are equally important to consider.* Recall that the mobile phase in RPLC typically includes water (20–80%) modified with polar organic solvents such as methanol, acetonitrile, or tetrahydrofuran. Benzene can interact with water through relatively weak dispersion and dipole–induced dipole interactions (the dipole of water inducing a dipole in benzene). Phenol, on the other hand, interacts not only through dispersion interactions but also through relatively strong dipole–dipole and hydrogen-bonding interactions. *Thus, the attractive forces between phenol and the mobile phase are considerably stronger than those between benzene and water.*

Putting all of this together, while phenol and benzene have roughly comparable attraction to the stationary phase, phenol is much more attracted to the mobile phase than is benzene. Therefore, phenol molecules, on average, spend a greater percentage of their time in the mobile phase compared to the stationary phase. Because total retention is simply the time spent in the stationary phase plus the time spent in the mobile phase, and because all solutes spend the same amount of time in the mobile phase (see the river analogy in Chapter 1), phenol is retained less than benzene because it spends less time in the stationary phase.

The above description of intermolecular interactions and their effect on relative retention can also be understood by recalling that retention is a thermodynamic process governed by the distribution constant, $K$ (also called the partition coefficient), for the equilibrium process $A_{mob} \iff A_{stat}$, where "$A$" represents an analyte. The stronger intermolecular interactions of phenol with the aqueous mobile phase shift the equilibrium toward the mobile phase relative to the same equilibrium for benzene. Thus, $K$ for phenol

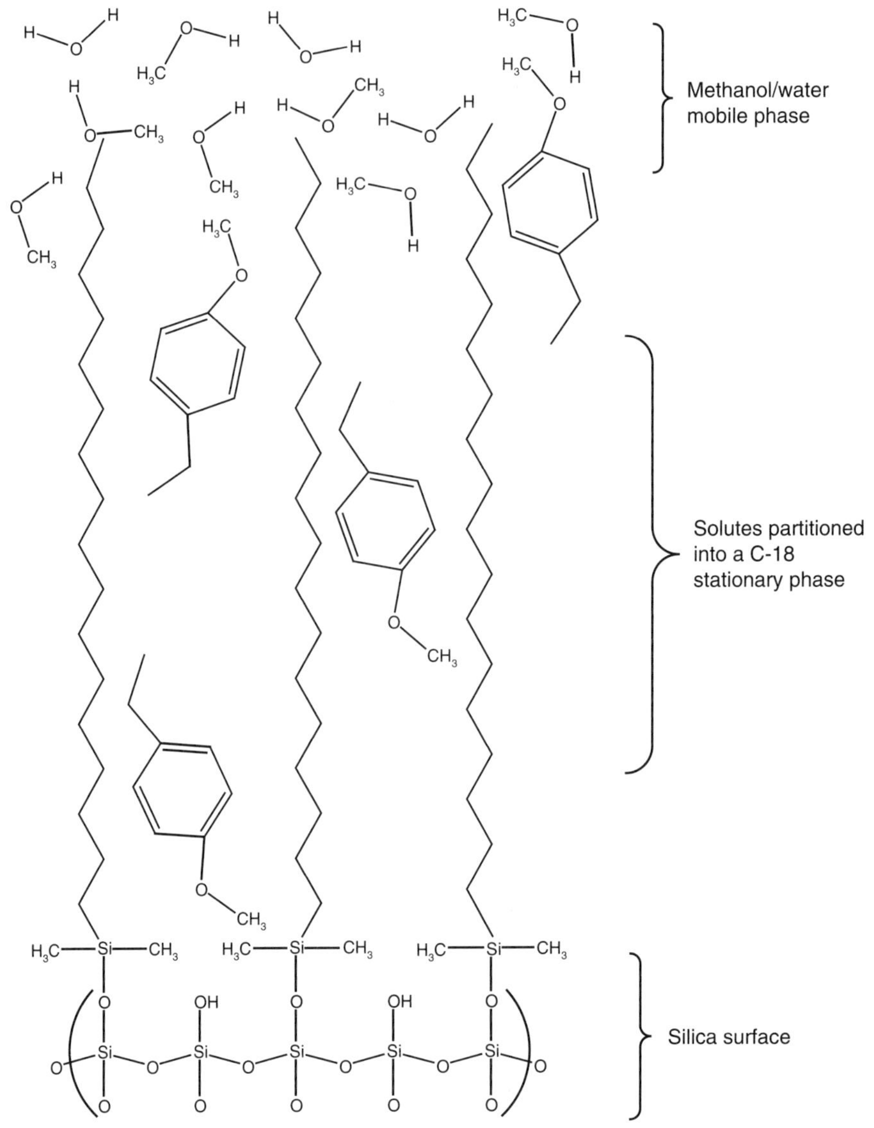

**FIGURE 3.7** Depiction of solute molecules (4-ethylanisole) partitioning into a C-18 stationary phase in RPLC.

is not as large as for benzene. Because the retention factor, $k$, is proportional to $K$, and is directly related to retention, it is easy to see that smaller equilibrium constants result in less retention. Therefore, phenol is less retained and consequently elutes earlier than benzene under typical RPLC conditions. It is important to note that this same analysis generally applies to the retention of all solutes: *polar solutes elute before nonpolar solutes in RPLC*.

### 3.4.2.1.  Removing Some Approximations. We began this section with the assumption that solutes partition entirely into the alkyl chain environment of the stationary phase.

While this is a useful approximation, it is not entirely correct. Because the alkane chains generally extend away from the surface of the particle and are modified by components of the mobile phase as shown in Figure 3.8, the stationary phase environment is not homogenous. It varies from the "bottom" where the silica support exerts some chemical influence, to the "top" of the alkyl chains that are in contact with the bulk mobile phase. Furthermore,

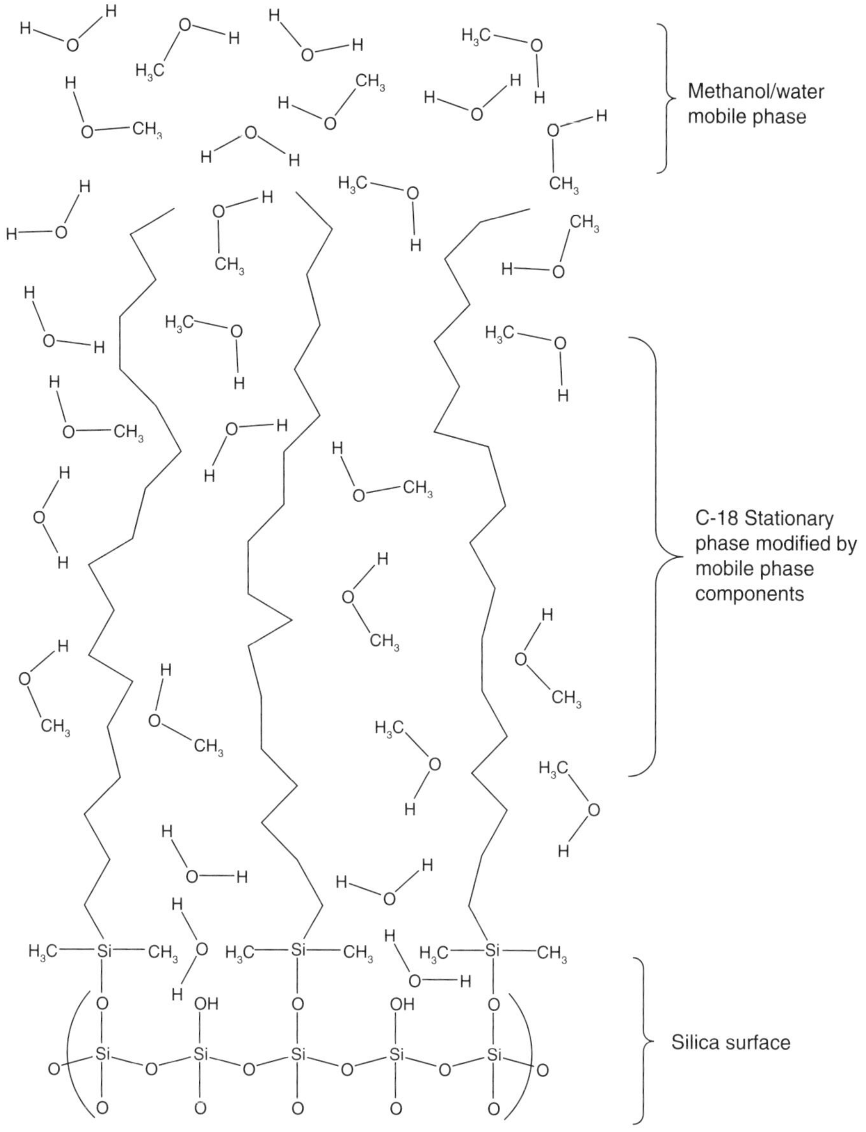

**FIGURE 3.8** Depiction of mobile phase modification of the C-18 chains in a typical RPLC column using a typical methanol/water mobile phase. While methanol is shown in this picture, all of the organic modifiers used in RPLC such as acetonitrile and tetrahydrofuran modify the stationary phase to some extent. Also notice that water is attracted to the polar silanol groups that may remain on the surface due to incomplete derivatization with C-18 chains.

the stationary phase environment changes as the composition of the mobile phase (e.g., methanol versus acetonitrile as modifiers) changes.

The nature of a solute also dictates where it resides when interacting with the stationary phase. Computational chemistry results obtained by Schure et al. suggest that nonpolar solutes are likely to be found both embedded within the alkyl chains and adsorbed across the top of them (i.e., adsorbed to the stationary phase).[17] Their work also shows that polar solutes such as 1-propanol preferentially reside near the top of the chains in the interfacial region between the stationary and mobile phases. They can also interact with silanol groups on the silica surface through hydrogen bonding. Furthermore, the distribution of 1-propanol throughout the stationary phase changes depending on the amount of organic modifier added to the mobile. The authors of this study used butane and 1-proponal as representative solutes, which can be viewed as alkyl relatives of benzene and phenol discussed above. Thus, one might expect similar behavior for benzene and phenol. So while viewing RPLC as a bulk two-phase partitioning process is a convenient first approximation and useful for predicting the order of elution of solutes, the actual retention of solutes depends on a range of subtle thermodynamic effects arising from the fact that the stationary phase is not homogeneous.

### 3.4.2.2. Stationary Phase Stability.
The chemical bond holding the stationary phase to the surface is important. The Si—O—Si bond is susceptible to hydrolysis at low pH. If hydrolysis occurs, the alkyl chain is cleaved from the silica particle, causing loss of stationary phase in the column. If the amount of stationary phase changes over time, retention is not reproducible and peak shapes degrade. Therefore, these phases should not be exposed to mobile phases with pH less than approximately 2. This, combined with the upper pH limit of approximately 8 dictated by the dissolution of the silica itself, sets the typical working pH range for RPLC. Some advances have been made to broaden this range, but it is wise to always check the manufacturer's guidelines regarding column stability so as to maintain reproducibility and maximize the life of the column.

---

**EXAMPLE 3.1**

Predict the order of elution of the following compounds in NPLC and RPLC: 1-butanol, octylbenzene, 3-nitrooctane, 3-nitroheptane.

**Answer:**
NPLC: octylbenzene, 3-nitrooctane, 3-nitroheptane, 1-butanol
RPLC: 1-butanol, 3-nitroheptane, 3-nitrooctane, octylbenzene

In RPLC, the small, polar, hydrogen-bonding 1-butanol interacts most favorably with water and therefore elutes first. The polar nitroalkanes elute next, with the shorter chain eluting earlier because it is less hydrophobic than the longer chain. Octylbenzene elutes last because it is favorably drawn into the stationary phase via interactions with the C-18 phase and because its interactions with water are weaker compared to the other compounds. NPLC will essentially be the reverse of this order, although this is not always the case.

**3.4.2.3.** *Monolithic Columns.* In contrast to conventional columns that contain discrete packing particles, monolithic columns are single, continuous solid rods containing relatively large channels called "through pores" in the 10–20 µm range and numerous smaller mesopores that have diameters of 1–5 µm.[18–21] The term "monolithic," from the Greek for "single stoned," reflects the solid, continuous nature of these columns. Figure 3.9 shows the through-pores and mesopores in a monolithic column. From a solute's perspective, the difference between conventional and monolithic columns is akin to moving through a tube packed with Styrofoam balls versus through a continuous cavern with interconnected channels.

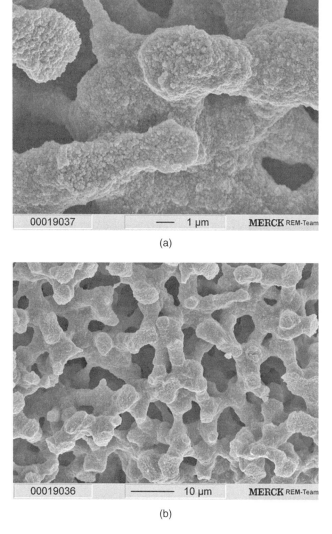

(a)

(b)

**FIGURE 3.9** Electron micrographs of the (a) macroporous and (b) mesoporous structures in a monolithic silica rod. (*Source:* Reprinted from *LC/GC*, 2001, 19(12), p. 1188 with permission from Dr. D. Lubda, and Merck KGaA, Germany.)

Both inorganic and polymeric monolithic columns are available. Inorganic columns are silica-based and modified with C-18 chains using chemistry comparable to that used to modify conventional silica particles. Thus, monolithic columns offer separation factors quite similar to conventional columns: the mesopores contain the majority of the surface area, having $300\,m^2$ of area per gram of material. Polymeric columns are made from polyacrylamides, polyacrylates, and poly(styrene-*co*-divinylbenzene). Compared to silica-based columns, polymeric columns suffer from swelling and shrinking depending on the solvent used and generally provide lower column efficiencies. They are, however, advantageous for bioseparations because they can be operated in a wide range of pH values and can therefore be cleaned using caustic mobile phases. For routine analyses, however, silica columns are more likely to be employed.

The main advantage of monolithic columns is that their open porous structure results in much lower back pressures when pumping solvents through the column. Thus, it is easier to achieve high flow rates without deteriorating the performance of the column. In addition, the slow adsorption and desorption kinetics that cause band broadening in conventional columns are accelerated in monolithic columns. Thus, the slope of the van Deemter plot at high flow rates is considerably smaller than in conventional columns, making the band-broadening behavior of monoliths comparable to that of 3.5 μm particles. Monolithic columns can be operated at flow rates as much as 5–10 times greater than traditional columns and thus are attractive options for high-speed/high-throughput applications.[18]

***3.4.2.4. Gradient Elution.*** Another factor when considering RPLC elution is the amount of time required to elute all components of a mixture. Consider a sample that contains both polar and nonpolar organic molecules. [Note: the term "polar," when applied to organic molecules, must be understood to mean polar *relative* to other organic molecules; organic molecules, even polar ones, still tend to be less soluble in water than they are in polar organic solvents such as ethanol or acetone.] In a mixture of polar and nonpolar solutes, the polar ones elute relatively quickly from the column because their intermolecular interactions help keep them in the highly aqueous mobile phase compared to the relatively nonpolar stationary phase. In order to separate these early-eluting organic components, it is necessary to use mobile phases that are mainly water, for example, 90% water and 10% acetonitrile. Such a highly polar mobile phase exploits any small differences in the hydrophobicity of the early-eluting compounds. If a highly aqueous composition is maintained, however, nonpolar solutes, which have very little affinity for water, remain in the stationary phase at the head of the column from the moment they first enter it. Because they seldom partition out of the stationary phase they never move down the column, never elute, and never get detected – or do so only over extremely long periods of time, resulting in broad flat peaks arising from diffusional broadening (see the discussion of band broadening in Chapter 1).

Thus, the challenge is to provide enough retention that early-eluting polar components are separated from one another, yet still achieve elution of nonpolar compounds in a reasonable amount of time. The solution is to systematically change the mobile phase composition during the course of the separation, and do so in such a way that the percent water decreases and the percent organic modifier increases. High percentages of organic modifiers such as methanol, acetonitrile, tetrahydrofuran, or isopropanol make the partitioning of nonpolar solutes into the mobile phase more favorable compared to their partitioning into highly aqueous phases. Thus, as the percentage of organic modifier increases, the solutes are less retained by the stationary phase. A typical example of a gradient profile is to start with an 80:20 water/acetonitrile mobile phase for approximately 2 or 3 min and then gradually change to a 20:80 water/acetonitrile mixture over a preset period of time. In this way, early-eluting (i.e., polar) compounds are exposed to the highly aqueous mobile phase while late-eluting species (i.e., nonpolar) are coaxed off the column with an organic-rich mobile phase. Throughout the gradient, then, a range of solutes of varying polarities can be separated due to the constantly changing nature of the mobile phase. This systematic change in the composition of the mobile phase is referred to as *gradient elution* and is commonly used in HPLC.[22] In contrast, it is sometimes desirable for practical reasons to keep the mobile phase composition the same throughout the entire analysis. When the mobile phase composition is kept constant, it is referred to as *isocratic* (i.e., same solvent) elution – a term first introduced by Csaba Horvath.[12,23]

An important side note: 100% water should never be used as an RPLC mobile phase because this causes changes inside the particles that are detrimental to column performance and that are difficult and time consuming to reverse. For this reason, RPLC mobile phases should always contain *some* (at least 5%) organic modifier.

### 3.4.2.5.  *Mobile Phase Solvent Strength.*

The "strength" of the mobile phase is a measure of its ability to elute solutes. Mobile phases that elute compounds quickly are "strong" compared to those that result in long retention times. In RPLC, highly aqueous mobile phases are "weak," whereas mobile phases containing a significant percentage of organic modifiers such as tetrahydrofuran or acetonitrile are "strong." Much effort has gone into characterizing the strength of mobile phases. Particularly noteworthy in this regard is the Snyder–Rohrschneider solvent selectivity triangle that groups solvents according to their ability to donate hydrogen bonds, accept hydrogen bonds, and to participate in dipole–dipole interactions.[15,24,25] A detailed analysis of their work is not required here, but it is important to be aware of it to fully understand liquid chromatographic separations. Consideration of Figure 3.10 illustrates the effect on solute retention of increasing the percentage of water in the mobile phase (i.e., decreasing the mobile phase strength) under isocratic elution conditions.

### 3.4.2.6.  *Mobile Phase Modifiers.*

As noted above, there are several modifiers used in RPLC; methanol, THF, acetonitrile, and isopropanol. These are commonly used because they have desirable characteristics for application in RPLC. Specifically, they are completely miscible with water, meaning that they dissolve in all proportions without phase separating, and they are generally transparent to UV–visible light. The miscibility issue is important because phase separation inside the column would lead to exceptionally poor chromatographic performance and potentially ruin the column, which costs hundreds of dollars each. Transparency to UV–visible light is important because many LC systems use UV–visible detectors. If the mobile phase absorbs UV–visible light, the background is exceptionally high, and solutes cannot be distinguished from the mobile phase as they elute. While it is most common to have binary mixtures for mobile phases,

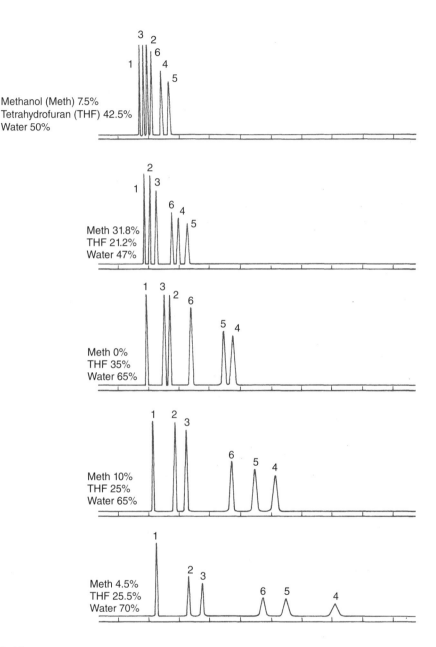

**FIGURE 3.10** Separations representative of five ternary mobile phases. Peaks: (1) benzyl alcohol, (2) phenol, (3) 3-phenylpropanol, (4) 2,4-dimethylphenol, (5) benzene, and (6) diethyl o-phthalate. (*Source:* After R.D. Conlon, "The Perkin-Elmer Solvent Optimization System" *Instrumentation Research*, p. 95 (March 1985). Used with permission from Perkin Elmer.)

combinations of three or more components are also frequently used. All of the separations (except the third) in Figure 3.10 use a ternary system. The specific components and their relative percentages can be adjusted to alter the separation factor ($\alpha$) in order to achieve the optimum resolution in the shortest amount of time.

**3.4.2.7. NPLC Mobile Phases and Gradients.** The same general considerations that dictate mobile phase components and gradients for RPLC also apply to NPLC. In NPLC, however, because the mobile and stationary phase polarities are opposite of those of RPLC, typical mobile phases are based on nonpolar solvents such as hexane and modifiers such as alcohols, THF, or dioxane are used to *increase* the polarity of the mobile phase, thereby making the mobile phase stronger. This is in contrast to RPLC where organic modifiers *decrease* the overall mobile phase polarity because they are being added to the highly polar solvent, water.

**3.4.2.8. Stationary Phases.** While C-18 stationary phases are by far the most commonly used RPLC columns, other phases that offer different retention characteristics and therefore different separation factors, are also commercially available. Common stationary phases are listed in Table 3.3 and depicted in Figure 3.11. Stationary phases made from shorter chains such as C-1 and C-4 retain solutes through a mechanism that is more similar to adsorption than true partitioning. Furthermore, the shorter chains do not produce as much retention as C-8 and C-18 columns.

**Amino and Cyano Phases.** Columns containing the polar amino and cyano silane derivatives offer the possibility of performing NPLC, but on bonded-phase particles. The advantage to this is that bonded phase particles offer better reproducibility and more robust separations than the bare silica, alumina, or titania columns typically used in classical NPLC. Furthermore, such phases offer different intermolecular interactions than do nonpolar phases and thus may be able to separate mixtures of compounds that cannot be separated on C-8 and C-18 phases.

**Polar-Embedded Phases.** Polar-embedded phases have a polar functionality as part of the stationary phase chain attached to the silica surface. Unlike traditional C-8 and C-18 phases, polar-embedded phases can be used in highly aqueous mobile phases.[26] Furthermore, the additional polar moiety introduces intermolecular interactions beyond those typically available in nonpolar phases. This changes the way some solutes are retained

**TABLE 3.3   Common LC Stationary Phases**

| Phase | Polar or nonpolar | Attributes |
| --- | --- | --- |
| C-1 | Nonpolar | Adsorption-type |
| C-4 | Nonpolar | Adsorption-type |
| C-8 | Nonpolar | Classic RPLC retention |
| C-18 | Nonpolar | Classic RPLC retention |
| Amino | Polar | NPLC-like retention |
| Cyano | Polar | NPLC-like retention |
| Phenyl | Nonpolar | RPLC with additional intermolecular interactions |
| Polar-embedded | Mixed | RPLC with additional intermolecular interactions |
| Di- and trifunctional | Nonpolar typically | Polymerized stationary phase with increased stability and reduced surface silanol activity |

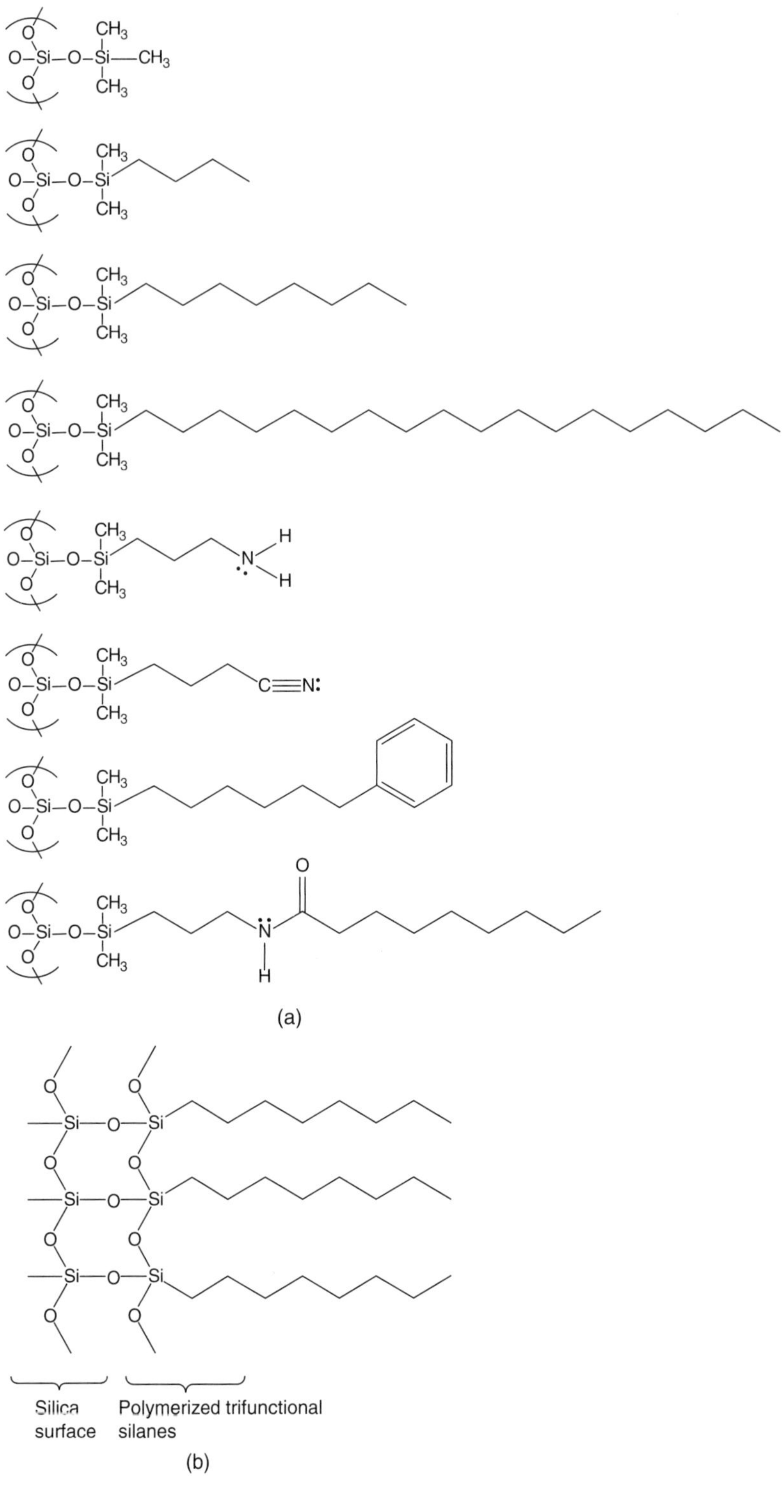

(a)

Silica    Polymerized trifunctional
surface   silanes

(b)

**FIGURE 3.11**  Depiction of common RPLC stationary phases. (a) From top to bottom: C-1, C-4, C-8, C-18, aminopropyl, cyanopropyl, phenylhexyl, and a polar embedded phase. (b) Horizontally polymerized trifunctional silanes.

and thus provides the possibility of separating some mixtures via RPLC that might not be easily separated on nonpolar phases.

**End-Capping and Di- and Trifunctional Silanes.** The stationary phase bonding chemistry shown in Figure 3.6 depicts the use of a monofunctional silane to attach C-8 chains to the silica particle. Unfortunately, this kind of stationary phase bonding leaves many of the surface silanol (Si—O—H) groups unreacted. Thus, while the stationary phase is intended to be completely nonpolar, the underlying surface is quite polar and capable of hydrogen bonding. In many applications, interactions of solutes with these unreacted silanols cause poor peak shapes and, in some cases, complete retention of solutes. This is particularly true with basic compounds that accept the hydrogen bonds donated by the silanol group. Because many pharmaceutical compounds are basic, the problems arising from residual silanol groups are particularly important to resolve if chromatography is to be used by drug companies to separate and quantify their formulations.

Because of the problems they cause, much effort is expended in trying to eliminate or mitigate the effects of surface silanol groups. One method for doing so is called endcapping, in which short chain (e.g., C-1) monofunctional silanes are reacted with the surface after the primary C-8 or C-18 chains have been attached. Because they are smaller, they can fit between some of the long chains and react with some of the residual silanol groups.

Another method for diminishing the effects of silanol groups is to use di- or trifunctional silanes to create the stationary phase (e.g., di- or trichloro alkylsilane).[27-29] In this case, the monomer that is being used to create the bonded phase is not only attached to the support particles but also the additional functionality makes it possible to polymerize the chains to one another as depicted in Figure 3.11b. This serves to make the stationary phase more robust. In other words, it is harder to remove from the surface and is therefore more resistant to removal from low pH mobile phases. Furthermore, it serves to mask the surface from the solutes and thereby diminish the negative effects of the surface silanol groups mentioned above.

The combination of mobile phase modifiers, stationary phases, and gradient elution provides great flexibility in separation conditions and makes the separation of most complex mixtures possible.

### 3.4.3. Ion-Exchange Chromatography (IEX)

Modern IEX was developed during the Manhattan Project in WWII, but the first commercial instrument to perform ion chromatography was introduced by Dow Chemical Company in 1975.[30-32] It was based on advances in ion-exchange stationary phases and a detection technique known as ion-suppression that were first described by Hamish Small and coworkers at Dow in the early 1970s.[32-34] In IEX, the stationary phase itself is charged to provide retention of charged analytes. This is accomplished using column packings that have charge-bearing functional groups covalently bonded to a polymer matrix, with associated counterions attracted through electrostatic interactions. The most common retention mechanism is thus the simple exchange of sample ions with the counterions as the sample moves through the column (see Figure 3.12).

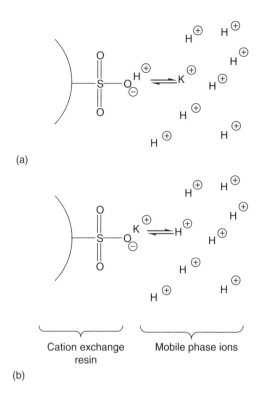

(a)

(b)

Cation exchange
resin

Mobile phase ions

**FIGURE 3.12**  Depiction of the retention of $K^+$ via cation exchange on a sulfonic acid exchange resin (strong acid type). The competing ion in this case is $H^+$, which is in the mobile phase. In (a), the eluent ion ($H^+$) is interacting with the phase, whereas in (b), the solute ion ($K^+$) is being retained. This equilibrium continues throughout the timeframe of the separation, with the solute moving down the column when it is displaced from the stationary phase and into the mobile phase by the eluent ions. Separation of solute ions (e.g., between $Ag^+$ and $K^+$) occurs because of their different affinities for the charged stationary phase and thus their different abilities to displace the mobile phase counterion from the stationary phase. Note: Other ions such as $Na^+$ or $Li^+$ are frequently used in place of $H^+$ as the eluent ion in the mobile phase.

*3.4.3.1.*  *IEX Phases.*  Some ion-exchange packings bear negatively charged groups and are used for exchanging cationic species. Others, designed for exchanging anionic species, contain positively charged groups. Figure 3.13 presents structures for some polymers commonly used in IEX chromatography. Sulfonate phases are used for cation exchange and quaternary amines are used for anion exchange. Sulfonate exchangers are exceptionally weak bases and quaternary amines are exceptionally weak acids, meaning that both groups remain in their charged form independent of mobile phase pH.

In contrast, the charge on some other exchangers, such as carboxylic acids and tertiary amines, is sensitive to pH. In these cases, the ability of the phase to retain solute ions depends on the pH of the mobile phase. If the pH is such that the majority of the phase is in its neutral form, retention of ions is slight. If the pH is adjusted such that the phase is totally charged, retention is at its maximum, provided that the analytes are completely and oppositely charged at the same pH.

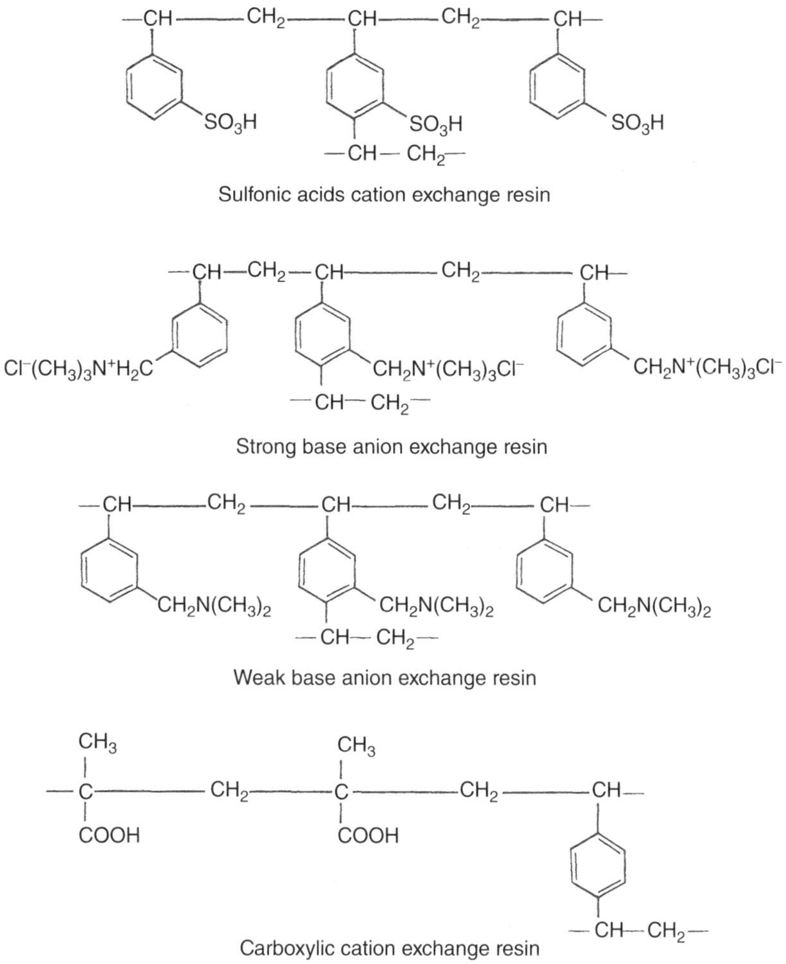

**FIGURE 3.13** Functional groups of ion exchange resins. (*Source:* From H.H. Willard, L.L. Merritt, Jr., J.A. Dean, F.A. Settle *Instrumental Methods of Analysis*, 7th edition, 1988, John Wiley and Sons, Inc. Reproduced with permission.)

***3.4.3.2. Exchange Equilibria.*** The primary process of IEX includes adsorption/ desorption of ionic materials in the mobile phase with an oppositely charged site on the stationary phase. For example, in the case of an ionized resin initially in the protonated form (symbolized as resin, $H^+$) in contact with a mobile phase containing potassium ions ($K^+$), an equilibrium exists

$$\text{resin, } H^+ + K^+ \leftrightarrow \text{resin, } K^+ + H^+ \tag{3.1}$$

which is characterized by the selectivity coefficient, $k_{K/H}$

$$k_{K/H} = \frac{[K^+]_r [H^+]_{mob}}{[K^+]_{mob} [H^+]_r} \tag{3.2}$$

where the subscript "r" refers to the resin phase and "mob" refers to the mobile phase. Strictly, the selectivity coefficient is constant only if the activity coefficient ratios in the resin and in the mobile phase are constant.

Retention differences are essentially governed by the physical properties of the solvated ions. The resin phase shows a preference for (1) ions of higher charge, (2) ions with smaller solvated radii (i.e., greater charge density), and (3) ions with greater polarizabilities. It is important to stress that it is the size and charge of the *solvated* ion that matters. In general, the solvated ionic radius limits the Coulombic interaction between ions, and the polarizability of the ions determines the van der Waals attraction. Together, these factors control the total energy of interaction between oppositely charged species and hence their overall retention on ion-exchange columns. So, while a lone gas-phase proton has a high charge density due to its exceptionally small size, protons in water (i.e., hydronium ions) are highly solvated, carrying with them a large sphere of water molecules that are attracted to the charge. As such, hydronium ions have very large *hydrated* radii and therefore low *charge densities*. This makes them only weakly attracted to a negatively charged stationary phase. For similar reasons, other small ions such as $Li^+$, $F^-$, and $OH^-$ tend to be weakly retained, whereas larger and more highly charged species tend to be strongly retained. For example, the retention of cations generally follows the order $La^{3+} > Al^{3+} > Ba^{2+} > Pb^{2+} > Sr^{2+} > Ca^{2+} > Ni^{2+} > Cu^{2+} > Zn^{2+} > Mg^{2+} > UO_2^{2+} > Ag^+ > Cs^+ > Rb^+ > K^+ > NH_4^+ > Na^+ > H^+ > Li^+$. For anions, the order from most to least retained is $C_6H_5O_7^{3-}$ (citrate) $> SO_4^{2-} > C_2O_4^{2-} > I^- > NO_3^- > Br^- > SCN^- > Cl^- > CH_3CO_2^- > F^- > OH^- > ClO_4^-$.[13,15]

Returning to the equilibrium above, the partitioning of potassium ions between the resin phase and the mobile phase can be described by means of a concentration distribution ratio, $D_c$

$$D_c = \frac{[K^+]_r}{[K^+]_{mob}} \tag{3.3}$$

Combining Equations 3.2 and 3.3 yields

$$D_c = k_{K/H} \frac{[H^+]_r}{[H^+]_{mob}} \tag{3.4}$$

This relationship shows that the concentration distribution ratio for *trace* concentrations of an exchanging ion is independent of the mobile phase concentration of that ion (neglecting activity effects). For example, as the mobile phase concentration of a trace ion doubles, so too does its concentration in the stationary phase, keeping the ratio of concentrations in the two phases constant.

However, the concentration distribution ratios are inversely proportional to the mobile phase solution concentration of the resin counterion, which is to be expected, because the counterion competes with the trace ion for exchange in the resin phase. Therefore, as more $H^+$ is introduced into the mobile phase, the additional $H^+$ competes with potassium ions for binding sites on the stationary phase, causing a net decrease in retention of potassium ions and hence faster elution times. In this way, the concentration of the mobile phase counterion can be used to control the retention times of analyte ions. Figure 3.14 illustrates the point

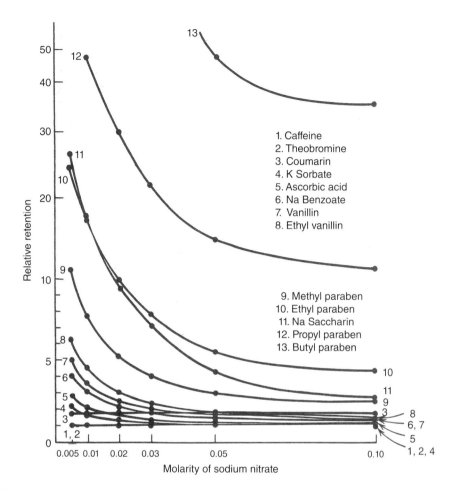

**FIGURE 3.14** Effect of mobile phase salt concentration on the retention of various food additives in ion-exchange chromatography. Column: $100 \times 0.21$ cm Zipax coated with 1% quaternary ammonium liquid-anion-exchanger; mobile phase: 0.01 M sodium borate (pH 9.2) with added $NaNO_3$ in water, 24 °C; 0.85 mL/min. (*Source:* Reprinted from J.J. Nelson, *J. Chromatogr. Sci.* 1973, 11, 28–35, with permission from Oxford University Press.)

that solute retention decreases as the mobile phase counterion concentration increases. In this figure, the stationary phase is an anion exchange phase. The mobile phase concentration of nitrate is systematically increased, which decreases the retention of anionic species.

To accomplish any separation of two cations (or two anions), it is necessary that one of these cations be taken up by the stationary phase resin in distinct preference to the other. In other words, the ions must have different strengths of interaction with the stationary phase. The strength of interaction is generally related to their difference in charge density (i.e., charge compared to the size of the hydrated radius of the ion). The differential ability of ions to compete with the mobile phase counterions for sites on the stationary phase and thus be retained is expressed by the separation factor, $\alpha$. Using potassium and sodium ions as an example yields

$$\alpha_{K/Na} = \frac{k_{K/H}}{k_{Na/H}} = k_{K/Na} \tag{3.5}$$

If the separation factor, $k_{K/Na}$, is unfavorable for the separation of potassium from sodium, meaning that both have similar affinities for the stationary phase, no variation in the concentration of $H^+$ in the mobile phase will improve the separation because the $H^+$ will displace each ion with nearly equal ease. The situation is entirely different if the exchange involves ions of different net charges. Here, the separation factor does depend on the concentration because, for electroneutrality reasons, multiple $H^+$ ions are required to displace polyvalent ions from the stationary phase, whereas only one is required to displace monovalent ions. In these cases, the more dilute the counterion is in the mobile phase, the more selective the exchange becomes for polyvalent ions. To understand this, it is useful to consider two extremes: a mobile phase swamped with $H^+$ and one with very dilute $H^+$. In the mobile phase that is swamped with $H^+$, the large excess of $H^+$ pushes the equilibrium in Equation 3.1 far to the left, meaning that the resin ionic sites are nearly completely occupied by $H^+$ ions out of sheer probability. In this case, no analyte ions, regardless of their charge or size, will be able to compete against the high concentration of $H^+$ for sites on the stationary phase. Essentially no analyte ions are retained, and hence there is *no* separation in this situation.

At the other extreme of dilute $H^+$, monovalent ions are more easily displaced from the stationary phase due to their lower charge and therefore lower columbic attraction compared to polyvalent ions. Furthermore, one $H^+$ can displace one monovalent ion. Thus, monovalent ions are more favorably displaced by $H^+$ than are polyvalent ions, which have stronger attraction to the stationary phase charged site and require multiple $H^+$ ions to be removed. In this case, polyvalent ions are preferentially retained over monovalent ions and the differential retention necessary to achieve the separation can be obtained by controlling the counterion concentration in the mobile phase.

### 3.4.3.3. *Detecting Ions in IEX.*

While general HPLC equipment has not been discussed yet, it is important to note here that IEX chromatography presents a challenge in terms of detecting non-UV absorbing species such as lithium, sodium, and potassium ions. One good detector for these types of species is a conductivity detector. In a conductivity detector, the eluent from the column flows through a cell with two electrodes, between which the conductivity of the eluent is measured. Ideally, as a band of analyte ions elutes, the conductivity of the solution in the cell changes due to the increased concentration of ions and gets recorded as a peak. However, this simple scheme suffers from the fact that the mobile phase in general often has a high conductivity because counterions are constantly pumped through the system to elute the sample ions. In particular, solutions containing $H^+$ and $OH^-$ ions have very high conductivities relative to solutions of other ions at the same concentration.[35] Therefore, the background signal is inherently high, and sample ion zones would be mere small perturbations on top of an already large signal. Analytically, it is never good practice to try to distinguish small differences in large signals if it can be avoided. For this reason, modern IEX performed with HPLC instruments uses ion suppression techniques to reduce or eliminate the high conductivity of the mobile phase prior to the eluent reaching the detector. It is in this area of ion suppression that Small and colleagues mentioned in the introduction to this section significantly advanced ion-exchange chromatography.[32-34] Different methods of ion suppression are explained in the next section.

**Two-Column Ion Suppression.** In this mode, the eluent from the separation column is passed through a second column before entering the detector. Suppose that nitrate, $NO_3^-$, is the analyte of interest and that sodium bicarbonate, $NaHCO_3$, is used in the mobile phase to facilitate elution of the analyte. As the $NO_3^-$ moves through the column, it displaces and is displaced by the $HCO_3^-$ ion. When the $NO_3^-$ finally elutes from the column as a solute band, for reasons of electroneutrality, its charge in solution is offset by $Na^+$ ions from the mobile phase. In the extreme, one can imagine the elution sequence as (1) pure mobile phase containing $Na^+$ and $HCO_3^-$ ions, (2) $Na^+$ and $NO_3^-$ in the mobile phase zone, and (3) back to pure mobile phase with $Na^+$ and $HCO_3^-$ ions. The task is to distinguish the conductivity of the analyte zone from the nearly equivalent conductivity of the mobile phase.

This can be achieved by passing the eluent through a second column. In this example, the second column is a cation exchange column, pretreated with acid such that it is in its fully protonated form. As the pure mobile phase passes through the column, sodium ions are exchanged for protons, which then protonate and thereby *neutralize* the bicarbonate ions as shown in Figure 3.15. In its neutral form, the resulting carbonic acid does not conduct electricity and is therefore not detected by a conductivity detector. Thus, the conductivity of the pure mobile phase (i.e., the baseline conductivity) is markedly reduced.

Now consider what happens as the analyte zone containing sodium and nitrate ions passes through the suppressor column. Again the $Na^+$ is exchanged for $H^+$, but because $HNO_3$ is a strong acid, the nitrate ions remain separate from the $H^+$ ions in the mobile phase. Due to the $H^+$ and $NO_3^-$ ions, this zone conducts electricity much better than the pure ion-suppressed mobile phase and is registered as a peak. As the solute zone leaves the detector cell, it is followed by more of the pure, ion-suppressed mobile phase and the detector signal returns to its low background signal.

The method for detecting cationic analytes is analogous except that an anion-exchange resin is used as the suppressor column in its fully hydroxylated form. In these cases, $HCl$ is often added to the mobile phase eluent and the excess $H^+$ is converted into $H_2O$ by the reaction with the $OH^-$ that is displaced by $Cl^-$ in the suppressor column. One drawback of this method is that the suppressor columns must be periodically returned to their fully protonated or hydroxylated forms by flushing them with strong acids or bases. This requires instrument downtime and therefore reduces analytical productivity.

Another method of ion suppression has been developed that acts on the same principles as above but does so in a continuous manner to avoid costly downtime. In this method, the eluent is passed down a core of a hollow-fiber cation (or anion) exchange membrane while a counter flowing stream of acid or base on the outside continuously supplies $H^+$ (or $OH^-$) to exchange for cations across the membrane. These systems use a high-capacity ion-exchange material and a void volume less than 50 µL. Figure 3.16 shows an expanded view of the suppressor, which sandwiches alternating layers of high-capacity ion-exchange screens with ultrathin ion-exchange membranes. The eluent flows lengthwise between the screens and the regenerant flows on both sides of the sandwich. Thus, interruptions for suppressor regeneration are eliminated. Ion-exchange sites in each screen provide a site-to-site pathway for eluent ions to the membrane. The high capacity of the suppressor unit expands the choice of eluents and permits the use of much higher eluent concentrations. This expands the range of analyte applications and permits gradient elution to be performed with ion chromatography.

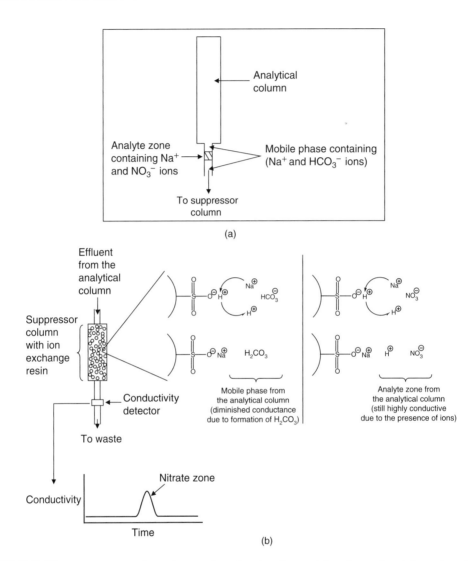

**FIGURE 3.15** (a) Depiction of the elution of a solute band containing the analyte (nitrate ions in this case). (b) Ion exchange inside the suppressor column reduces the conductivity of the mobile phase but not that of the analyte (nitrate) zone, leading to a chromatogram of conductivity versus time.

***3.4.3.4. IEX Uses.*** IEX is the dominant method for anion analysis because these species are difficult to analyze using other instrumental methods. Thus, IEX is very important for the analysis of species such as nitrate, phosphate, sulfate, perchlorate, fluoride, chloride, and bromide in environmental and waste water samples. IEX chromatography is also used to analyze inorganic metallic ions such as $Na^+$, $K^+$, $Li^+$, and $Ca^{2+}$ that are medically important, and $Ba^{2+}$, $UO_2^{2+}$, and others of importance to the nuclear industry. Some metallic ions are of great importance to the computer industry in which trace metallic impurities can cause significant problems in the performance of computer chips and hard drives. Still other ions of importance include pharmaceutical compounds (many of which

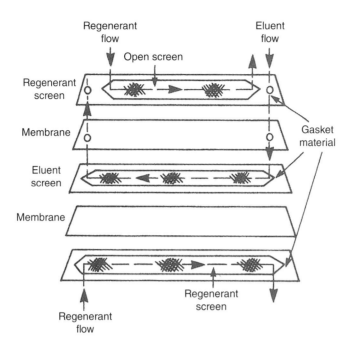

**FIGURE 3.16** Expanded view of the micromembrane suppressor for ion exchange chromatography. (*Source:* Used with permission from Thermo Fisher Scientific, the copyright owner.)

are amine bases and can therefore be protonated), food additives, dyes, amino acids, DNA, RNA, nucleotides, and a wide range of environmental pollutants (particularly those found in water). IEX finds particularly widespread use in proteomic studies and in protein purification methods. Because of all of these different applications, IEX is an important and widespread analytical technique.

### 3.4.4.  Hydrophilic Interaction Chromatography (HILIC)

While RPLC is capable of separating a wide range of compounds, small polar molecules typically elute quickly because of the high water content used in the mobile phases. Such compounds often overlap or coelute. HILIC is an alternative mode for separating polar and charged compounds.

HILIC, like NPLC, uses polar stationary phases such as bare silica or short-chain amino and cyano phases. The mobile phase typically contains at least 2.5% water and more than 60% of an organic solvent, with acetonitrile being common. In RPLC, increasing the water content of the mobile phase typically causes an increase in the retention of organic solutes. Under HILIC conditions, the opposite behavior occurs.

The acronym "HILIC" was first used in an article by Alpert in 1990.[36] In that article, he showed that the retention of peptides, amino acids, and other polar biological molecules decreased as the percent organic modifier in the mobile phase increased when using very low percentages of water and a polar stationary phase – behavior that is opposite to that under typical RPLC conditions.

The mechanism underlying HILIC has been the subject of much investigation since the first report by Alpert. Several modes of retention seem to be active in HILIC, namely, partitioning, adsorption, and ion exchange.[36,37] The relative contribution of each mode depends heavily on the nature and charge of the solutes, the stationary phase structure, and the characteristics of the mobile phase, such as water content, organic modifier, pH, and ionic strength.

The partition mechanism arises from a layer of water that is adsorbed to the particle surface throughout the pores (see Figure 3.17).[38] In the case of bare silica particles, the Si—O—H groups attract and hold water molecules from the mobile phase to the surface through hydrogen-bonding. This creates a liquid layer near the pore surface that is richer in water than is the mobile phase. Polar molecules are attracted to this layer and are thus retained in a partition-like mechanism (i.e., solutes partition out of the bulk mobile phase and into the liquid layer at the surface). The more hydrophilic a solute is, the more it favors partitioning into the immobilized water layer, and thus the more it is retained. Separation of different types of solutes is then based on their relative hydrophilicities. Solutes that can hydrogen bond can also interact directly with the silanol groups on the surface, in what might be seen as adsorption to a specific site on the surface, as opposed to general partitioning into the water layer.

Ion-exchange mechanisms also can contribute to HILIC separations. Bare silica has a $pK_a$ of around 4.5, such that at pH values near and above this value, the surface carries a negative charge due to deprotonation of the Si—O—H groups. These charged sites can act as cation exchange sites. Amino ($—(CH_2)_3NH_2$) phases are positively charged at pH values

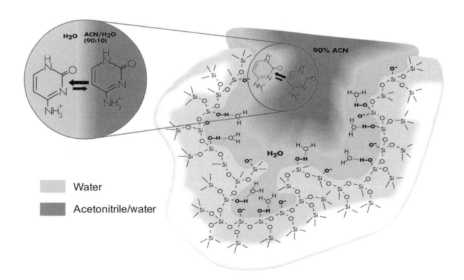

**FIGURE 3.17** Depiction of the water-rich layer on the surface of the pores in bare silica particles in HILIC. This image also depicts two of the different retention modes; partitioning of the solute from the organic-rich phase into the water-rich layer, and ion exchange (specifically cation exchange in this instance). (*Source:* U.D. Neue, *Separation Sci.*, Vol. 2(10), 2010. Reproduced with permission of Separation Science.)

below their $pK_a$ and thus provide anion-exchange sites. Zwitterionic phases and mixed mode phases also exist, providing complex blends of multiple retention mechanisms. Naturally, the pH of the mobile phase and that of the local environment in the aqueous layer near the particle surface influence the ionization state of the solutes and the surface. This, in turn, significantly influences retention. Likewise, the addition of salts into the mobile phase changes the ionic strength and also impacts retention.

Much remains to be understood about HILIC, but its popularity as a separation mode, particularly for polar molecules that elute quickly and are therefore not easily separated by RPLC, is growing rapidly. Another reason for the interest is that the organic-rich mobile phases used in HILIC are quite volatile compared to RPLC mobile phases. This makes HILIC especially attractive for coupling to mass spectrometric (MS) detection. For example, biochemical and pharmaceutical samples frequently contain polar and/or charged species, and because mass spectrometry is useful for studies in these fields, interest in HILIC–MS has grown rapidly.

### 3.4.5. Size Exclusion Chromatography (SEC)

NPLC, RPLC, ion-pair, IEX, and HILIC all take advantage of the ability of molecules to participate in intermolecular interactions to achieve separations. Differences in the molecular and ionic structures of solutes lead to differences in the strengths of interactions and thus cause some molecules and ions to be more retained than others, and thus separated from one another. Another difference in molecules that can be exploited to achieve separations is simply differences in their size. Separations based on the size of molecules are referred to as size exclusion chromatography (SEC) or gel permeation chromatography (GPC).[39,40]

SEC is a noninteractive mode of separation typically used to separate macromolecules and polymers ranging in molecular weight from 1000 to over 500 million Da. The particles of the column packing have variously sized pores and pore networks and act like a molecular maze for the solutes to pass through. Solute molecules are retained in or excluded from the pores based of their hydrodynamic molecular volume – that is, their size and shape. Strictly speaking, separation in SEC is not based on the molecular weight of the species; however, molecular weight and molecular size are often correlated.

As the sample passes through the column, the solute molecules are sorted by the pores in the packing material inside the column (see Figure 3.18). Very large molecules cannot enter many of the pores. Thus excluded, they travel mostly around the exterior of the packing and elute at the void volume of the mobile phase and are essentially unretained. Very small molecules diffuse into most or all of the pores. Because they explore a large volume of space within the column, small molecules exit the column last. Between these two extremes, intermediate-size molecules penetrate some passages but not others and are consequently slowed in their progress through the column and exit at intermediate times.

Naturally, the terms "large," "small," and "intermediate" are relative to the diameters of the pores in the packing material. The range of compounds with sizes that fall into the "intermediate" category, which can therefore be separated using SEC, can be deliberately selected by choosing particles with the appropriate pore size distribution.

Mobile phase flow

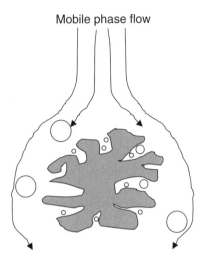

**FIGURE 3.18** Depiction of size exclusion chromatography. The shaded area represents a porous particle. Small circles represent small solutes that can enter all pores, medium circles represent solutes that can enter only some of the pores, and large circles represent solutes that are too large to enter any of the pores and thus are not retained. Mobile phase flow carries the solutes around the porous particles. If the particles are of appropriate size, they can diffuse into pores. If they are too large, they continue past the particles and elute at the dead time.

### 3.4.5.1. *Retention Behavior in SEC.* The essential behavior of a solute and the characteristics of porous column packings can be discussed in simple terms. For a packed column of porous particles, the total bed volume, $V_t$, is given by

$$V_t = V_m + V_s + V_g \tag{3.6}$$

where $V_m$ is the void volume of the mobile phase (i.e., the unbound solvent in the interstices between the solvent-loaded porous particles) estimated by the elution of a totally excluded solute; $V_s$ is the cumulative internal volume within the porous particles and available to a totally included solute or molecule of solvent (also denoted as $V_i$); and $V_g$ is the volume occupied by the solid matrix.

Under typical SEC conditions, the retention (elution) volume, $V_R$, of a solute is the volume of effluent that flows from the column between the time when the sample is injected and its emergence in the effluent, that is,

$$V_R = V_m + K V_s \tag{3.7}$$

where the distribution constant, $K$, represents the fraction of the internal pore volume accessible to the solute. Rearranging Equation 3.7 yields

$$\frac{V_R - V_m}{V_s} = K \tag{3.8}$$

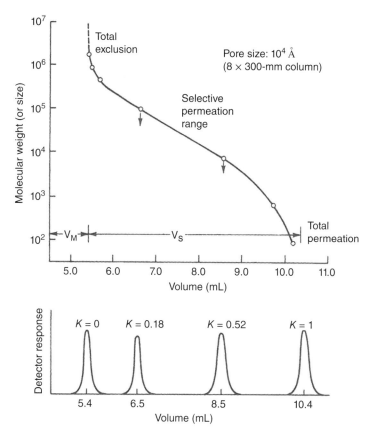

**FIGURE 3.19**   Retention as a function of solute size in SEC. (*Source:* H.H. Willard, L.L. Merritt, Jr., J.A. Dean, F.A. Settle *Instrumental Methods of Analysis*, 7th edition, 1988, John Wiley and Sons, Inc. Reproduced with permission.)

Totally excluded molecules elute in a single void volume; in this case, $V_R = V_m$, which yields $K = 0$. For small molecules that can enter all the pores of the packing, $V_R = V_m + V_s$, and hence $K = 1$. Intermediate size molecules elute between these two limits and $K$ ranges from 0 to 1. An elution graph is shown in Figure 3.19. The upper portion is the graph of the logarithm of molecular weight of solutes versus retention volume. It is a sigmoidal curve in which there is a linear range of effective permeation between the limiting values that correspond to total exclusion ($K = 0$) and total permeation ($K = 1$). The *exclusion limit* of a packing is the size above which molecules will not be retained (and therefore not separated). Similarly, the *permeation limit* is the size below which solutes will not be separated due to total inclusion in all pores. These two parameters are used to characterize SEC columns.

   The total number of sample components that can be resolved is limited for any particular packing. However, as detailed below, particles of various pore sizes are commonly available. When the probable molecular dimensions or weights of the sample components are known, selecting the particular pore size or exclusion limit of the packing is usually straightforward.

**TABLE 3.4    SEC Polymeric Particle Pore Sizes and the
Molecular Weight Range They Can Separate**

| Pore size (Å) | MW range |
|---|---|
| 50 | 100–10,000 |
| 500 | 1,000–70,000 |
| 1,000 | 40,000–5,000,000 |

***3.4.5.2.  SEC Column Packings.*** SEC particles are commonly made from polymeric cross-linked polystyrene/divinylbenzene. The pore size is controlled by varying the degree of cross-linking between the polymer chains. Table 3.4 shows average pore sizes for a polymeric SEC resin and the MW range of the proteins they separate.[40]

Polymeric particles can swell in different solvents, and high pressures can cause them to be compressed, so analysts must be aware of and understand the operating characteristics of SEC columns. Porous silica particles provide a more rigid structure that can be used at greater pressures and does not suffer from solvent swelling. They do suffer, however, from having a chemically active surface. The silanol moieties (Si—O—H) can retain solutes via dipole–dipole, dipole–induced dipole, and hydrogen-bonding interactions, introducing a second separation mechanism. This mixed-mode retention (intermolecular interactions with size exclusion) can result in poor peak shapes, incomplete separations, and complicated data interpretation. Deactivating the surface with silylating agents can help reduce these problems but cannot eliminate them completely.

***3.4.5.3.  SEC Uses.*** While SEC is a valuable tool for separating large molecules and finds widespread use in polymer analysis and biochemistry, it does not provide the same type of resolution generally observed in HPLC. As a rule of thumb, sample components need to vary in MW by about 10% in order to be completely resolved. This, combined with factors such as polymer swelling and mixed-mode retention, limits the applicability and power of SEC separations. However, for some applications such as protein and polymer analysis, it is the perfect tool and therefore needs to be considered and understood.

### 3.4.6.  Affinity Chromatography

Thus far we have seen how separations can be achieved by taking advantage of *nonspecific* intermolecular interactions and the size of solutes. Affinity chromatography is based on the *specific binding* of substrates to enzymes, often referred to as "lock and key" binding due to the specific "fit" of the solute in the enzyme.[41,42] The general scheme of affinity chromatography involves the covalent attachment of an immobilized biochemical (called an affinity ligand) to a solid support. When a sample is passed through the column, only the solute that selectively binds to the ligand is retained; the other sample components elute without retention. The retained solute molecules are then eluted from the column by changing the mobile phase conditions. The major advantage of this technique is its tremendous specificity, which permits rapid isolation of a solute in good yield such that it can be collected and used in subsequent studies.

***3.4.6.1. Affinity Column Matrices.*** Column matrices for affinity chromatography encompass a wide variety of materials: organic gels such as beaded agarose, cellulose, dextran, polyacrylamide, and combinations of these polymers. The matrix should be sufficiently hydrophilic to avoid nonspecific binding of solutes and stable to most water-compatible organic solvents. A porous matrix permits a high degree of ligand substitution and is more accessible to larger molecules. Preactivated agarose supports are commercially available. Affinity ligands such as antibodies, enzyme inhibitors, or enzymes themselves are covalently linked to the support matrix to make columns that selectively bind the complementary molecule.

Figure 3.20 shows the general scheme behind affinity chromatography. Ideally, only those solutes that specifically bind to the modified surface are retained while all other sample components elute without being retained. The mobile phase composition is then changed to elute and subsequently quantify the retained analyte. Retained solutes may be eluted either by adding excess ligand to the mobile phase or by adding a compound that interferes with the solute ligand interactions. In biospecific elution, the mobile phase modifier (called the inhibitor) is a free ligand similar or identical to the immobilized affinity ligand or the solute. The inhibitor competes for sites on the ligand or solutes. For example, if immobilized glucosamine (a small molecular substrate) were used to purify a lectin (a protein), either glucose or $N$-acetyl-D-glucosamine, which compete with glucosamine for binding to lectin, might be used as the inhibitor.

Nonspecific elution involves the denaturation of either the ligand or analyte by means of pH, chaotropic agents (i.e., those that disrupt molecular structures such as KSCN or

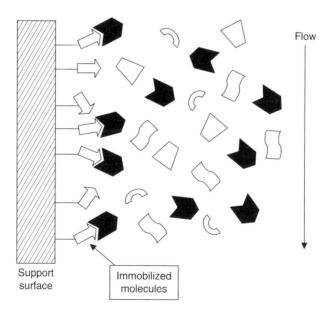

**FIGURE 3.20** Depiction of affinity chromatography. Only those solutes that are complementary to the molecules immobilized on the stationary phase are strongly retained. In this depiction the solid black "solutes" are complementary to the immobilized arrows. All other solutes ideally pass through the column unretained.

urea), organic solvents, or by changing the ionic strength of the mobile phase, and serves as a more general method of elution.

---

**EXAMPLE 3.2**

Which mode of liquid chromatography would be best for separating the following?

(a) The products of a reaction involving common organic compounds
(b) The components in lactated Ringer's solution
(c) A mixture of *o*-, *m*-, *p*-aminobenzoic acid
(d) The chain size distribution in a cellulose sample
(e) The insulin receptor in liver cell extracts
(f) A mixture of peptides and other small, polar biomolecules.

**Answers:**
(a) (a) RPLC is an appropriate choice for most general separations of organic compounds.
(b) Lactated Ringer's solution is an electrolyte solution that is isotonic with blood and is used in hospitals and animal research. Because it contains electrolytes such as sodium, potassium, calcium, and chloride, IEX is appropriate.
(c) NPLC often provides good separations of positional isomers, although in many cases RPLC can also be used.
(d) Cellulose is a polymer, so SEC is appropriate.
(e) The insulin receptor is a protein and could be separated from other proteins in the extract using affinity chromatography with insulin covalently linked to the stationary phase.
(f) HILIC could be considered in this case because retention increases with the hydrophilicity of the compound. Small polar compound may elute too quickly and thus not be resolved if RPLC is used.

---

## 3.5. HPLC INSTRUMENTATION

The fundamental principles upon which molecules can be separated were described in the previous sections. Now the task is to understand how modern HPLC instruments are configured such that they convert the fundamental principles and theories into actual separations from which analytical information is gained.

A schematic of a typical HPLC instrument is shown in Figure 3.21. The solvent reservoirs hold the mobile phase components. HPLC systems commonly have one to four solvent reservoirs. Systems with two reservoirs are referred to as "binary." "Ternary" and "quaternary" are used to describe systems with three and four reservoirs, respectively. Mobile phase solvents must be filtered and degassed before reaching the column. Degassing is necessary because small bubbles caused by dissolved gases such as $O_2$ and $N_2$ can lodge in the detector cell, causing erratic baselines and distorted solute peaks. In modern instruments, degassing is achieved either by sparging the solvents with helium

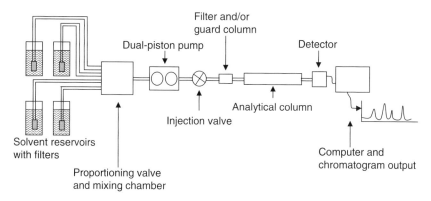

**FIGURE 3.21**  Schematic of a liquid chromatography system.

or by pulling the dissolved gasses out of the solvent through narrow, gas-permeable tubes through which the solvent flows on its way to the column. Filtering is necessary so that small particles do not get transferred to the column where they lodge at the head of the packing and restrict the flow of solvent through the column. If this occurs, poor peak shapes, irreproducible flow control, and exceptionally high pump pressures may result. Filtering is usually accomplished by a filter cartridge placed between the mobile phase reservoir and the rest of the system. Additional filters prior to the column may be added to eliminate the complications associated with particles clogging the column. Both degassing and filtering can also be done "off-line" using vacuum systems to pull solvents through filters. It is simply more efficient and convenient to have the instrument perform these functions in line with the rest of the system if such an option is available.

### 3.5.1.  The Proportioning Valve

The electronically controlled proportioning valve is responsible for creating the desired mobile phase compositions (usually specified electronically by the user). It is here that, for instance, a 50:50 methanol/water mix at a total flow rate of 1 mL/min is created. In this case, only 0.5 mL/min of each solvent is needed for the pump to deliver 1 mL/min. Thus, the total draw from the pumps is "proportioned" between the different reservoirs to achieve the desired mobile phase composition and flow rate.

### 3.5.2.  Mixing Chamber

Solvents are often passed through a mixing chamber prior to being pumped through the column. The resulting homogeneous mixture assures that all of the analyte molecules experience the same mobile phase composition as they pass through the column.

### 3.5.3.  Pumps

Reciprocating pumps are by far the most commonly used pumps in HPLC instruments. A schematic of a reciprocating pump is shown in Figure 3.22. In a reciprocating pump,

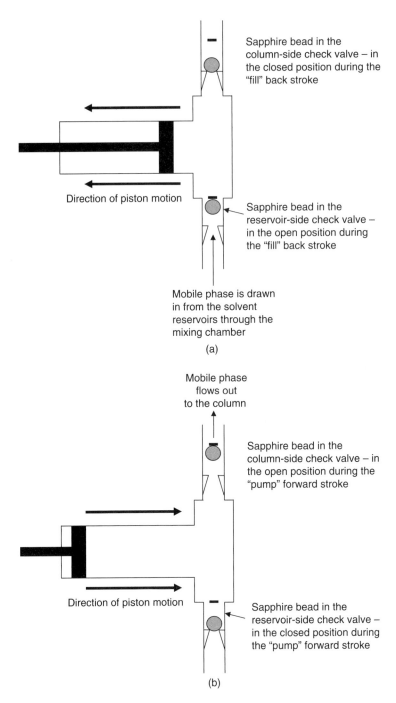

Sapphire bead in the column-side check valve – in the closed position during the "fill" back stroke

Direction of piston motion

Sapphire bead in the reservoir-side check valve – in the open position during the "fill" back stroke

Mobile phase is drawn in from the solvent reservoirs through the mixing chamber

(a)

Mobile phase flows out to the column

Sapphire bead in the column-side check valve – in the open position during the "pump" forward stroke

Direction of piston motion

Sapphire bead in the reservoir-side check valve – in the closed position during the "pump" forward stroke

(b)

**FIGURE 3.22** Operation of reciprocating pumps. (a) The pump fills with mobile phase during the back stroke as solvent is drawn in from the solvent reservoirs and through the proportioning and mixing valves. (b) The mobile phase is pushed out to create flow through the column during the forward stroke of the piston.

a motor-driven piston moves back and forth to first fill the pump chamber with mobile phase and then to force the mobile phase through the column. The heart of the reciprocating pumps is the check valves that regulate the direction of flow as the pump operates. As the piston is moved backward in the chamber, two floating beads or balls, one in each check valve, react to the pull created by the backward-moving piston. The solvent reservoir side is designed such that the ball, which would otherwise be drawn into the piston chamber, is blocked from doing so. Note, however, that the path to the mixing chamber (and ultimately the solvent reservoir) is open, and the solvents are drawn into the pump chamber as the piston moves backward. At the same time, the ball in the check valve on the column side of the pump seals off the column from the pump as the piston moves backward. This prevents the mobile phase that is already in the column from being drawn backward into the pump.

As the piston is moved forward and the mobile phase is pushed out of the chamber, the ball on the solvent reservoir side is driven down, where it seals off the pump from the solvent reservoirs. In this way, the solvent that was just drawn in is not allowed to flow back into the reservoirs. Instead, as the piston moves forward, the ball on the column side moves to a barrier and the mobile phase flows onto the column.

The cycle is then repeated with backstrokes drawing solvents from the reservoirs into the system and forward strokes delivering the solvents to the column. All the while, the proportioning valve distributes the total amount of "pull" created by the backstrokes among the various solvents so that the amount of each solvent drawn into the system creates the desired mobile phase composition. The mobile phase flow rate through the column is controlled by regulating the rate at which the pistons cycle through this process. As you increase the flow rate on an HPLC instrument, you can hear the increase in the frequency of the "whine" produced by the motors that are driving the pistons.

One drawback to this type of discontinuous pumping is that it produces pulses of mobile phase. These pulses can themselves produce background signals that create bumpy baselines that complicate the measurements of peak heights and areas. To minimize this effect, many HPLC systems use a dual-head pump arrangement. In these systems, two pumps are operated 180° out of phase such that one pump is being filled while the other is delivering solvent to the column and vice versa. Thus, there is a continuous flow of solvent to the column, which reduces the severity of the pulses. In addition to dual-head pumps, it is common for HPLC systems to use pulse dampers. These are chambers that take up some of the pulsation energy and thereby decrease the severity of the pulses before the mobile phase is delivered to the column. Animations of this entire process are available on the Internet.

Other types of pumps such as screw-driven syringes and pneumatic amplifier pumps exist but do not find widespread use. Good descriptions of the operation of these other pumping systems are available.[13,15,43]

### 3.5.4.  Injection

***3.5.4.1.  Manual Injections.*** Manual injections in HPLC are performed using rotary valves and sample loops (see Figure 3.23). In the "load" position, mobile phase from the pump passes directly onto the column. In this position, a syringe is used to fill the sample loop with the sample. The sample loop is simply a hollow tube with a fixed length

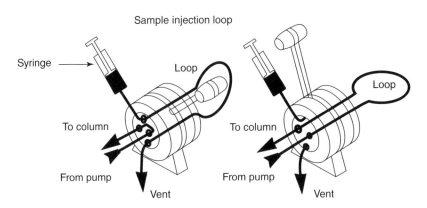

Sample injection loop

**FIGURE 3.23**   HPLC manual sample injection loop. Left: Load position in which the syringe is used to fill the sample loop. Note that the mobile phase from the pump goes directly into the column, bypassing the sample loop in this position. Right: Injection position in which the mobile phase flow from the pump flows through the sample loop, pushing the sample onto the column. (*Source:* From H.H. Willard, L.L. Merritt, Jr., J.A. Dean, F.A. Settle *Instrumental Methods of Analysis*, 7th edition, 1988, John Wiley and Sons, Inc. Reproduced with permission.)

and therefore a fixed volume. Sample loops typically range in volume from 10 to 100 µL. The volume injected into the loop is intentionally greater than the loop's internal volume to ensure that the loop is completely filled with sample. The excess volume simply exits through a waste vent. In the same way that a glass pipet is rinsed multiple times with the solution to be pipetted, it is wise to fill the sample loop multiple times to ensure that the composition of the mixture in the loop accurately reflects the actual sample composition. Residual solvent on the walls of the loop can dilute the sample if this precaution is not heeded.

To perform an injection, the valve is rotated to the "injection position." This diverts the mobile phase through the sample loop, which pushes the sample onto the column. The valve remains in the injection position for the remainder of the analysis. At the end of the analysis, the valve is rotated back to the load position so that a new analysis can be performed. Because the loop is now full of mobile phase, it is again wise to flush it multiple times with the sample before performing the next injection.

Using sample valves and loops ensures that the same volume is delivered every time. In this regard, injection in HPLC is much more reproducible (~0.1% rsd) than in GC (up to 1% rsd). Nevertheless, as in GC, using internal standards can significantly improve the accuracy of analyses (see the discussion of internal standards in Chapter 2). In addition, the continuous flow of mobile phase though the sample loop during the analysis reduces the possibility of sample carryover.

***3.5.4.2.   Autosamplers.***   Most modern HPLC instruments are equipped with autosamplers. These systems can contain up to 100 sample vials that can be analyzed automatically. A specified vial is picked up by a robotic arm and transported to an injection needle. During the sampling process, mobile phase is diverted away from the needle using automated valves. A specified amount of sample (typically 1–10 µL) is drawn up into the needle. The vial is returned to its original location and the sample needle, which now contains the sample, is placed in line with the column. A valve is automatically turned to direct the mobile phase flow through the needle and onto the column. In this way, the sample plug is pushed out of the needle and onto the column.

### 3.5.5.  The Column and Particles

Most analytical columns are made of stainless steel to resist the high pressures (500–3500 psi) exerted when trying to pump mobile phase through the packed column. Standard diameters range from 1 to 5 mm. Columns with 4.6 mm diameters have historically been the most common, but smaller diameter columns are rapidly becoming the norm. The narrow-bore columns require less solvent to achieve the same linear velocities (i.e., same analysis time), offer increased sensitivities, and are easier to couple to mass spectrometer detectors because smaller amounts of solvent need to be removed compared to larger-bore columns. Most modern HPLC systems are compatible with column diameters down to 2.1 mm. Narrower columns require modified injectors, detectors, and pumps.

Typical HPLC columns range from 3 to 30 cm in length. Shorter HPLC columns are used for fast analyses, while longer columns offer increased theoretical plates that can be advantageous for resolving nearly overlapping peaks. Column lengths are ultimately limited by the pressures required to push mobile phase through them. As pressures increase, porous packing particles can be crushed and this destabilizes the particle bed, leading to poor chromatographic peak shapes and a column that must be discarded. This can be expensive because HPLC columns range in price from $200 to $800.

As discussed earlier, typical packing particles are spherical porous silica and range in size from 1 to 10 μm in diameter, with 3 and 5 μm particles being common, although sub-2 μm particles are also popular for many applications. Smaller particles offer significantly enhanced column efficiencies (increased $N$) and produce very narrow peaks and better resolution, thereby allowing higher flow rates and ultimately faster analyses. Smaller particles pack together more tightly, requiring significantly increased pressures to achieve reasonable flow rates. These increased pressures can require special pumps and the narrow peaks can require faster responding detectors. Also as discussed above, the silica particles are highly porous, with pore sizes ranging from 60 to 300 Å for typical NPLC and RPLC applications. These interconnected pores significantly increase the surface area of the particle. Thus, the vast majority of the bonded stationary phase is actually inside the particle as opposed to the outer surface of the particle. The mobile phase does not actually flow through the pores, rather, solutes diffuse into the pores where they are retained. The solutes move down the column only as they diffuse out of the pores and back into the mobile phase flowing past the particle surface.

A popular alternative to totally porous particles is superficially porous particles (SPP), also called core–shell particles. Such particles are made of a solid silica core surrounded by a porous layer (the shell). Figure 3.24 shows depictions of SPPs. Several manufacturers offer such particles, with diameters typically ranging from 2 to 5 μm (as shown in the figure), but sub-2 μm SPPs are available.

These particles are receiving a lot of attention because they offer comparable performance to sub-2 μm particles. More specifically, because of improved mass transfer

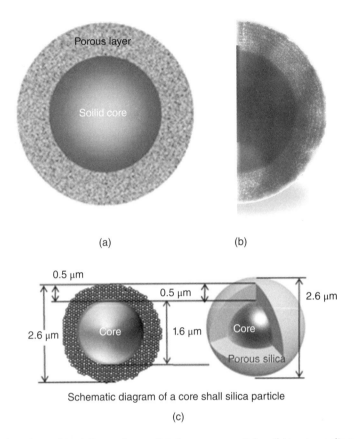

Porous layer

Solid core

(a)                                                  (b)

0.5 µm

0.5 µm                                             2.6 µm

2.6 µm          Core          1.6 µm        Core

                                              Porous silica

Schematic diagram of a core shall silica particle

(c)

**FIGURE 3.24**    A variety of depictions of superficially porous particles. (b) is a tunneling electron micrograph of a cross-section through a 2.6 mm Kinetex core–shell particle from Phenomenex. (a) (*Source:* Reproduced with permission of SIELC Technologies.) (b) (*Source:* Reproduced with permission of Phenomenex.) (c) (*Source:* Reproduced with permission of Nacalai USA.)

(i.e., diffusion) effects that result from the nature of the particles, their band-broadening behavior is similar to that of much smaller particles, with very little increase in the C-term in the van Deemter equation with increasing mobile phase velocity (see Figure 3.25).[44] This means that the mobile phase velocity can be increased significantly, without significant increases in band broadening that would otherwise ruin resolution. This makes it possible to do fast separations. While sub-2 µm particles also allow for faster separations, they require higher pumping pressures that necessitate special pumps. Because SPPs are larger, they do not cause the same high back pressures as sub-2 µm particles and therefore they can be used with standard HPLC equipment.

***3.5.5.1. Column Temperature Control.*** It is important to remember that retention in chromatography is ultimately a thermodynamic process and is therefore temperature dependent. Thus, control of temperature is important for obtaining reproducible retention times. Many HPLC systems have a thermostatted column compartment to maintain a constant column temperature, usually between 25 and 40 °C. For faster analyses, operating

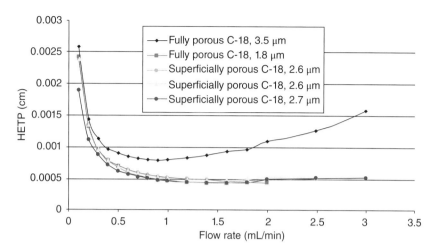

**FIGURE 3.25** van Deemter curves comparing fully porous particles to superficially porous particles (SPPs). Note that the SPPs have essentially the same performance characteristics as the sub-2 μm particles. This is particularly important in the high flow rate region (i.e., high linear velocities) as it enables high-speed LC with the larger SPPs without sacrificing resolution and without the high back pressures encountered with smaller particles. (*Source:* Reproduced with permission, Courtesy of CHROMacademy and Agilent Technologies, Inc. © Agilent Technologies, Inc.)

at elevated column temperatures (75–200 °C) is often required. This can require special column heaters and also requires that the mobile phase be preheated before entering the column. If the mobile phase is not preheated, the lack of thermal equilibrium between mobile phase and column temperature can lead to poor chromatographic performance because the mobile phase is essentially warmed by the column as the separation is occurring.[45]

### 3.5.5.2. Other Particle Materials.
While silica is by far the most common solid support material, titania and zirconia are also sometimes used. Zirconia in particular offers some advantages for high-temperature, high-speed separations. In addition to these inorganic-based particles, polymeric beads are also used, especially in systems where small particles (1 μm in diameter) are desired. Finally, as noted earlier, application of special columns may be required for techniques such as SEC, where the mechanism of separation is not adsorption or partitioning but rather based on the physical size of the analytes.

### 3.5.6. Guard Columns

The packing materials in analytical columns are held in the column by frits at the front and back of the column. These frits are essentially very fine wire meshes. These meshes can become clogged with small particles that are pumped through the system with the mobile phase. If this occurs, the entrance to the column is essentially blocked and the back pressure required to force the mobile phase through the column skyrockets. At this point, the column is essentially ruined and must be discarded. To prolong the life of analytical columns, guard columns are often inserted ahead of the analytical column where they act as both physical and chemical filters. Guard columns are relatively short (usually 5 cm) and contain a stationary phase similar to that in the analytical column. They protect the analytical column by trapping particulate contamination that arises from the mobile phase or from degrading components such as pump and injector seals before they can reach the analytical column. A guard column also extends the lifetime of the expensive separation column by capturing strongly retained sample components, preventing them from contaminating, and therefore permanently modifying, the head of the analytical column. Guard columns also protect the analytical column by saturating the mobile phase with any stationary phase or solid support material that might dissolve in the mobile phase. For example, silica dissolves at high pH. So if a mobile phase of pH 8 or 8.5 is being used, it would eventually dissolve some of the analytical column particle packing, which would ruin the column. If a guard column is in place, the guard column particles dissolve instead, and saturate the mobile phase with silica. Now when the mobile phase enters the analytical column, no more silica can be dissolved due to saturation of the mobile phase with the dissolved silica from the guard column. The expensive analytical column is thereby preserved at the sacrifice of the less expensive guard column.

Guard columns are not an unmitigated blessing, however. The increased system volume introduces some extra-column band broadening that can jeopardize the resolution of some separations. If, however, some resolution can be sacrificed without seriously degrading the separation, the use of guard columns (or in-line filters at a minimum) is highly recommended.

### 3.5.7. Detectors

As described in the earlier sections, a lot of time and energy is spent preparing and pumping the mobile phases in proper compositions that allow analyte molecules to be separated in cleverly designed and well-constructed columns. This would all be for naught if the molecules could not be detected as they elute from the column. Several different detection schemes exist, all of which take advantage of some physical or chemical attribute of analytes to sense their presence.

There are several desirable properties for a detector. First, detector flow cells are low volume so that they do not produce significant band broadening that degrades the resolution achieved in the analytical column. They also have rapid response times in order to avoid peak distortions or completely missed signals. For rapid HPLC analyses, this means detectors have time constants of 0.3 s or less in order to capture the narrow solute zones produced by high-efficiency columns.[46] Finally, the sensitivity of the detector should match the demands for detecting the analytes at the concentrations present in the samples. This last requirement demands high sensitivity detectors for trace analysis.

***3.5.7.1. Ultraviolet–Visible Absorbance Detectors.*** A schematic of a flow cell used with absorbance detection methods is shown in Figure 3.26. Optical absorption detectors operate on the fact that many molecules absorb UV–visible electromagnetic radiation. In fact, UV–visible detectors are by far the most commonly used detectors for HPLC, although mass spectrometers are also common. Basically, two different types

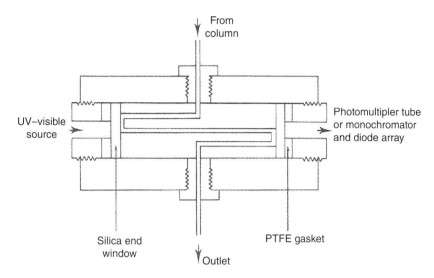

**FIGURE 3.26**  Flow cell for UV–Visible detection in HPLC. (*Source: High Performance Liquid Chromatography*, by S. Lindsay, D. Kealy (Eds.) 1987. Copyright John Wiley and Sons, Inc. Reprinted with permission.)

of absorption detectors are available, fixed-wavelength and multiwavelength detectors. Commercially, multiwavelength diode array instruments are considerably more common than fixed-wavelength instruments today; however, older instruments may have fixed-wavelength detectors and thus is it necessary to understand the capabilities of both detection methods.

**Fixed-Wavelength Detectors.** A fixed-wavelength detector uses a light source that emits maximum light intensity at one or more discrete wavelengths that are isolated by appropriate filters. The light passes through a flow cell containing the column effluent. Quartz windows allow the light to be transmitted through the cell to the detector. Such detectors are inexpensive but suffer from limited flexibility. For example, a mercury lamp typically allows detection only at 254, 280, 313, 334, and 365 nm. Many aromatic compounds have absorbance bands that encompass 254 nm and thus can be detected, although not maximally, at that wavelength. Some instruments have multiple filters on a wheel so that the specific detection wavelength can be changed throughout the separation to maximize the signal for the various sample components. Fixed-wavelength detectors produce chromatograms that are plots of absorbance versus time. The only information that can be gained from such a chromatogram is that a compound that absorbs a specific wavelength of light eluted at a specific time. Solute identification is performed by injecting known compounds and comparing retention times. Quantification is performed by injecting standard solutions of known concentrations and creating calibration curves. The use of internal standards leads to measurements with lower uncertainties than do analyses without internal standards. To help achieve better sensitivity, it is common for flow cells to have a "Z-configuration" such as that shown in the figure. This increases the path length that the light travels through the column effluent. This helps increase the signal because absorbance is proportional to path length (recall the Beer–Lambert law).

**Multiple-Wavelength Detectors.** The photodiode array detector (PDA), also referred to as a diode array detector (DAD), is a very popular detector for HPLC systems. In these systems, both a tungsten-halogen lamp and a deuterium lamp produce continuous emission throughout the UV and visible spectral ranges. The light is passed through the flow cell and then separated into its constituent wavelengths that are sensed by a detector array. In this way, the *entire* UV–visible spectrum of the content in the flow cell is recorded in as little as 0.01 s. This recording of the spectrum continues throughout the analysis such that at each instant of the separation, the entire UV–visible spectrum is available. At the start of the analysis, when only mobile phase is flowing through the flow cell, the UV–visible spectrum is recorded. Each subsequent spectrum is compared to the original. In this sense, the original spectra acts like a "blank." When only mobile phase is eluting, the recorded spectrum is virtually identical to the original, so there is no net difference at any wavelength. When a solute is in the flow cell, however, its absorption properties are likely different than the mobile phase and the UV–visible spectrum of the solute is recorded.

The PDA allows for greater flexibility and sensitivity for a wider range of compounds than do single wavelength detectors. Because an entire spectrum is recorded several times a second, multiple chromatograms can be produced. For example, with software, one can see the chromatogram that would result by plotting the absorbance at a specific wavelength, for instance 220 nm, versus time. The software would then use each collected spectrum to select just the absorbance at 220 nm in every spectrum and plot it versus time. The same could be done for any wavelength from 190 to 600 nm (the operating range of most PDA systems). In this way, the wavelength(s) that generates the largest signals for the desired compounds can be selected. This can significantly enhance the sensitivity of the method.

Diode array detection is also beneficial in that the entire UV–visible spectra of the sample components are recorded and can therefore be used for identification purposes by comparing the recorded spectra to spectral databases. Figure 3.27 illustrates how the different spectra of the compounds can be visualized. The differences in the spectral characteristics of the four main components are quite clear, and these differences can help identify the compounds.

Furthermore, PDA detectors allow for purity analyses to be conducted. If two peaks slightly overlap, the area of overlap produces a different UV–visible spectrum compared to nonoverlap regions. In this way, comparisons of the recorded spectra from various points under a given peak can be used to determine if the peak represents a single component or if it actually arises from two or more components that are coeluting.

While photodiode array detectors are more expensive than single-wavelength detectors, their increased flexibility, sensitivity, and information content makes the moderate increase in cost justifiable for many applications. Scanning instruments in which a grating is rapidly scanned to collect the spectra of eluting compounds, or to quickly raster to discrete wavelengths, are also available. These detectors are not nearly as common as diode arrays, however, because they offer no significant advantages.

**Air Bubbles.** As mentioned earlier, bubbles can be formed in the chromatographic system (transfer lines, pump, column, detector, etc.) by dissolved gases in the mobile phase solvents. Under high pressures, these bubbles are small, but in the detector, which is near atmospheric pressure, they can expand and become quite large relative to the size of the detection window. As the bubbles pass the window, or get trapped in the flow

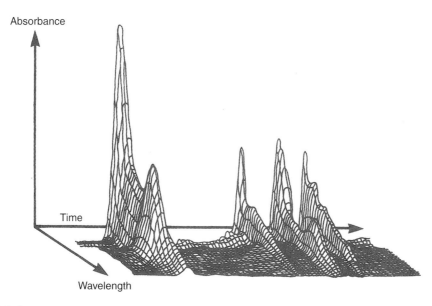

**FIGURE 3.27** Absorbance data as a function of wavelength and time acquired using a diode array detector. (*Source:* From H.H. Willard, L.L. Merritt, Jr., J.A. Dean, F.A. Settle *Instrumental Methods of Analysis*, 7th edition, 1988, John Wiley and Sons, Inc. Reproduced with permission.)

cell, they deflect light from the lamps and thereby create spikes in the chromatograph or noisy backgrounds. To eliminate the potential for bubbles to form, it is common to degas solvents. This can be done manually by drawing a vacuum over the solvents or sparging with helium. Most LC instruments have in-line vacuum degassing that eliminates the need to degas the solvents prior to use.

**3.5.7.2. Fluorescence Detectors.** Fluorescence detection is another important detection scheme used in HPLC systems. Just as in UV–visible detection, in a fluorescence detector the column effluent is passed through a flow cell. Light in the UV–visible region of the spectrum is continuously passed through the flow cell. Solutes in the flow cell can absorb the light and subsequently reemit it, usually at a longer wavelength. This emitted light, known as fluorescence, is collected, usually at 90° to the excitation beam, and passed to a detector to generate the analytical signal.

Not all molecules absorb UV–visible radiation and even fewer reemit the absorbed energy as light. The likelihood that a molecule will fluoresce generally increases with the number of rigid aromatic rings in the molecule and with increasing conjugation. Polyaromatic hydrocarbons (PAHs) are an example of this class of molecules. The requirement for fluorescent solutes limits the number of applications for fluorescent detectors. However, when the analytes of interest are fluorescent, fluorescence detection is exceptionally more sensitive than the other common HPLC detection schemes.

Many commercial HPLC fluorescence detectors can operate in a fixed excitation and emission mode or can scan both the emission and excitation monochromators in order to collect emission or excitation spectra as the solutes pass through the flow cell. For solute identification purposes and to maximize sensitivity, the capability to control both the emission and excitation monochromators is particularly useful. For simple quantitative measurements based on calibration curves, simply collecting fluorescence at a single wavelength may be all that is required. If very intense excitation is required to detect very low levels of solutes, lasers, instead of typical sources such as xenon lamps, can be used to induce fluorescence.

As was said earlier, few molecules fluoresce, so fluorescence detection is not as widely used as UV–visible absorption detection. To overcome this limitation, the analytes can be derivatized with fluorescent tags. This is particularly important in the detection of biomolecules such as DNA and proteins. Detection limits down to the nanogram level have been achieved using fluorescence detection for biological compounds. Thus, the coupling of HPLC with fluorescence detection provides a powerful and sensitive method for separating and quantifying important biomolecules in complex mixtures such as cell extracts, serum samples, and urine specimens.

### 3.5.7.3. *Refractive Index Detectors.* While UV–visible and fluorescence detectors operate on the principle of absorption of electromagnetic radiation by the solute molecules, refractive index detectors (RIDs) respond to differences in the refractive index of the pure mobile phase compared to when solutes are present in the mobile phase.

A *deflection-type* RID, which measures the deflection of a beam of monochromatic light by a double prism, is shown in Figure 3.28. Eluent passes through half of the prism while pure mobile phase passes through the other half. An optical mask confines a beam of light from a tungsten lamp to the face of the sample and reference compartments. The beam, collimated by a lens, passes through the compartments and is reflected back by the mirror through the compartments again. The beam is then focused on a beamsplitter before passing into twin photodetectors. When pure mobile phase is flowing through both the sample and reference compartments, a "zero adjust" or "null balance" is used to adjust the beams such that the same signal is generated by both detectors (i.e., they are balanced). If the refractive index of the mobile phase changes due to the presence of a solute, the beam from the sample compartment is slightly deflected. As the beam changes location on the detector, an out-of-balance signal is generated that is proportional to the concentration of the solute.

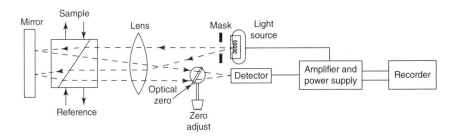

**FIGURE 3.28**  A deflection-type refractometer. (*Source:* Reproduced with permission of Waters Associates.)

RI detectors are linear over a wide range of analyte concentrations. They are also particularly advantageous for detecting solutes that do not absorb UV–visible light and which, therefore, cannot be detected by UV–visible detectors. Important compounds such as carbohydrates, some polymers, and aliphatic compounds fall into this category. However, RI detectors are highly temperature sensitive because refractive index is a function of temperature. Thus, the temperature must be tightly controlled to provide stable backgrounds. RI is less sensitive than UV–visible detection, but if the sample does not absorb UV light, some sensitivity is better than none.

### 3.5.7.4. *Mass Spectrometry Detectors.*

Mass spectometry (MS) detectors are based on the fact that molecules can be ionized and subsequently detected as they elute from a column. Recent advances in LC columns, pumps, ionization interfaces, and mass analyzers have made the combination of liquid chromatography and mass spectrometry relatively inexpensive, easier to operate, and quite powerful as an analytical tool. Furthermore, MS detectors are universal in that charge can be imparted to any analyte of interest and subsequently detected by a MS detector. This universal nature adds to the flexibility and general utility of LC/MS as an analytical tool.

The main tasks in interfacing LC to MS are (1) eliminating the mobile phase so that the MS detector is not overwhelmed by background signal, (2) ionizing the analyte molecules in the gas phase, and (3) achieving the low pressures required by MS despite comparatively high pressures that exist at the column outlet.

These tasks are performed by two commonly used ionization interfaces: (1) electrospray ionization (ESI) and (2) atmospheric pressure chemical ionization (APCI).

**Electrospray Ionization (ESI).** In 2002, Professor John Fenn of Virginia Commonwealth University shared the Nobel Prize in Chemistry with Koichi Tanaka of Shimadzu Corporation and Kurt Wüthrich of the Swiss Federal Institute of Technology for their contributions to the development of electrospray ionization (ESI). This critical development in mass spectrometry allows large biomolecules to be multiply charged in a soft ionization process. This technique overcomes the challenges associated with MS analysis of molecules that are difficult to volatilize and ionize while keeping their structures intact. In this way, compounds such as proteins can now be analyzed by mass spectrometry. Furthermore, ESI can be interfaced with other instrumental techniques such as capillary electrophoresis and liquid chromatography. Thus, the electrospray interface allows separation techniques to be coupled to the power of mass spectrometry.

The ESI interface (Figure 3.29) creates solvent-free analyte ions in three steps: (1) nebulization and charging, (2) desolvation, and (3) ion evaporation.[47] In the nebulization and charging step, the HPLC column effluent flows through a narrow needle that nebulizes the liquid into tiny droplets, which are then charged by passing through a semicylindrical electrode held at a high potential relative to the needle.[48] The potential difference between the needle and the electrode imparts charges to the surface of the droplets. After the droplets form, heated nitrogen flowing past the nebulizing needle evaporates the mobile phase solvent in the droplets. As the droplets shrink, they reach a size at which the Coulombic repulsion between the charges on the surface exceeds the cohesive forces holding the droplets together. The droplets explode and the process repeats itself until the analytes are solvent-free and left with residual charge from the electrospray

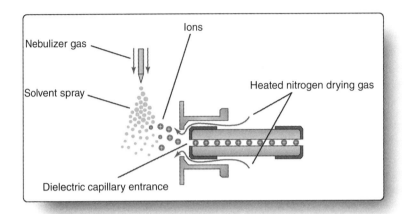

**FIGURE 3.29** Schematic of an electrospray ionization (ESI) interface. Droplets in the solvent spray acquire a charge from an electrode. Eventually the charge is transferred to solutes through a process of solvent evaporation and droplet explosion. Charged solutes are drawn through the capillary sampling orifice and into the mass analyzer. (*Source:* © Agilent Technologies, Inc. 2001 Reproduced with Permission, Courtesy of Agilent Technologies, Inc.)

process. These gas-phase analyte ions are attracted to a sampling capillary or cone by the low pressure that exists on the mass spectrometer side of the sampling orifice. Once through the sampling orifice, solute ions are introduced into one of any number of mass analyzers, including quadrupoles, triple quadrupoles, time-of-flight (TOF), Orbitraps, and ion traps. While ESI can be used for analytes that produce singly charged ions, it is particularly useful for analyzing solutes that can hold multiple charges such as proteins, peptides, and oligonucleotides. Because of the multiple charges, molecules with molecular weights up to 150,000 Da can be accurately analyzed.

**Atmospheric Pressure Chemical Ionization (APCI).** APCI is another commercially available LC/MS interface (Figure 3.30). Again the process starts with nebulizing the column effluent by flowing it through a small needle around which flows a nebulizing gas such as nitrogen to aid the nebulization process. The nebulizing chamber is heated to temperatures between 250 and 400 °C, which evaporates the spray droplets, leaving mobile phase and analyte molecules in the gas phase (but uncharged as of yet). A corona discharge needle is used to ionize the solvent molecules, which transfer their charge to gas phase analyte molecules. The analyte ions are then drawn into the sampling capillary by the low pressure of the mass selector and focused by ion optics on the entrance to the mass selector.

APCI is generally limited to analytes of moderate molecular weight because multiply charged analytes are less common with this type of ionization. The same mass analyzers mentioned above for ESI interfaces can be used with APCI interfaces.

**Information Gained by LC/MS.** The requirement of the mass spectrometer is that it be fast enough to collect mass spectra in a short time (typically under 2 s). In this way, the entire mass spectrum of the column effluent can be collected many times each minute. A chromatogram in LC/MS, as in GC/MS, is typically a plot of the total ion current (TIC) arising from all of the ions striking the MS detector versus time. As an analyte zone elutes from the

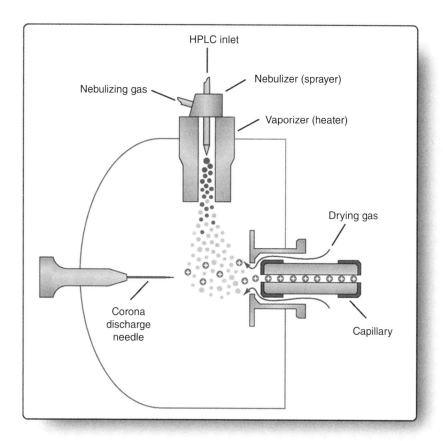

**FIGURE 3.30** Schematic of an atmospheric pressure chemical ionization (APCI) interface. Droplets are ionized by the corona discharge needle. Charge is transferred from the solvent to solute molecules which are eventually drawn into the sampling capillary and passed into the mass analyzer. (*Source:* © Agilent Technologies, Inc. 2001 Reproduced with Permission, Courtesy of Agilent Technologies, Inc.)

column, the number of ions created increases and thus the signal rises in proportion to the number of analyte ions present. In this way, standard solutions and the peak areas arising from them can be used to establish calibration curves for quantitative analyses.

The power of LC/MS, however, lies in the fact that the entire mass spectra of all of the analytes in the sample mixture are recorded as they elute. These mass spectra provide valuable information regarding the molecular weights of the analytes, as well as fragmentation patterns that can help identify the compounds. In this way, the MS detector acts similarly to the DAD except that a solute's mass spectrum provides a more definitive confirmation of solute identity than does its UV–visible spectrum. The combination of diode array followed by mass spectrometry detection, however, provides a very powerful analytical technique that can separate components of a complex mixture, provide nearly unambiguous identification of those components, and quantify the components through the use of standards and calibration curves.

In addition, by operating in selected ion mode (SIM), the detector can be set to register only particular mass-to-charge ($m/z$) ratios. To use SIM to test for a certain compound (e.g., a well-known doping agent used by athletes) in a complex mixture that will yield multiple overlapping peaks, the MS is set to a mass-to-charge ratio that is specific to the target analyte (in this example the doping agent) and which is not produced by any of the other sample components. In this mode, the detector becomes specific for compounds with the $m/z$ ratio of interest and does not detect any other components, eliminating the potential for overlapping signals and significantly simplifying the chromatogram. SIM mode also has the benefit of offering significantly lower detection limits.

**High Resolution Mass Spectrometry (HRMS).** While there are several different types of mass analyzers that can be coupled to LC systems, there is considerable interest in those that can detect very small mass differences – in other words, those with high resolution. The two primary types in this regard are time-of-flight (TOF) mass analyzers and the Orbitrap mass analyzer introduced in 2005.[49] The Orbitrap mass analyzer, for example, offers mass resolution on the order of 100,000, meaning it can discriminate between two particles with a nominal mass of 200 amu that differ by just 0.002 atomic mass units.

The Orbitrap has an inner electrode and outer electrode (see Figure 3.31) that trap ions between them. The ions oscillate along the long axis of the analyzer with a frequency that depends on their mass-to-charge ratios, creating an image current. The image current is converted to a frequency spectrum using a Fourier transformation.

Because they have such high resolution, as well as high mass accuracy, the molecular weights measured by HRMS instruments can often be used to identify compounds. The resolution also helps separate ions that have nearly identical mass-to-charge ratios, as shown in Figure 3.32.[50] The power of HRMS is heightened considerably when coupled to a HPLC or UHPLC system. The chromatographic system separates compounds and the HRMS analyzes their mass-to-charge ratios as they elute from the column. This can frequently

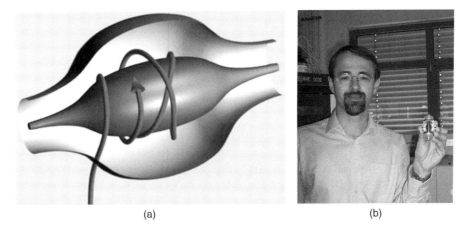

(a)                                                                              (b)

**FIGURE 3.31** (a) Depiction of an Orbitrap mass analyzer and the oscillatory nature of an ion with it. (b) Alexander Makarov holding his invention. (*Source:* Photograph used Courtesy of Thermo Fisher Scientific; copying prohibited.)

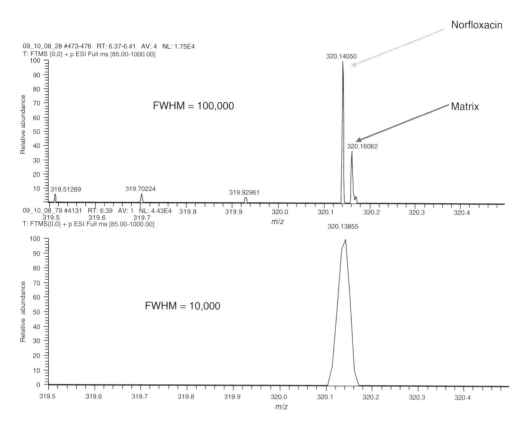

**FIGURE 3.32** Example of the advantage of high-resolution mass spectrometry to discriminate a target molecule from a matrix component of equivalent nominal mass. With lower resolution (10,000), the presence of norfloxacin in the sample could not be confirmed, but high resolution (100,000) makes this possible. (*Source:* Reprinted with permission from *Anal. Bioanal. Chem.*, 2012, 403, 1233–1249. Copyright 2012, Springer.)

lead to nearly unambiguous identification of the important compounds in biological samples, drug samples, food extracts, petroleum products, and a whole host of other complex matrices.

**Tandem Mass Spectrometry.** While HPLC–MS is powerful, it can be even more powerful when tandem mass spectrometry detection is used. In tandem mass spectrometry, symbolized as MS/MS, the ions from a first mass analyzer are further fragmented and the fragments sent to a second mass analyzer. Different ions, even if they have highly similar nominal masses, are unlikely to fragment in identical ways. So the second mass analyzer provides further discrimination of two closely related compounds. The combination of UHPLC with MS/MS is currently used by regulatory labs to rapidly screen food products for toxic and prohibited substances and to screen biological samples such as urine and blood from athletes for banned performance-enhancing drugs, as discussed in the introduction to this chapter.

### 3.5.7.5. *Evaporative Light Scattering Detectors.*

Another universal detector, but one that does not find widespread use, is the evaporative light scattering detector (ELSD) (Figure 3.33). In an ELSD, the column effluent is nebulized by passing it through a narrow needle while mixing it with a flow of nitrogen gas. The solvent is subsequently evaporated in a drift tube. This is usually aided by heating the drift tube, although commercial models with ambient temperature evaporation capabilities are available for the analysis of volatile and thermally labile components that could be evaporated or degraded by high temperatures. After the volatile mobile phase solvent components are vaporized, fine analyte particles remain and pass through a detection zone. In the detection zone, the particles are passed through a laser beam and scatter the laser light. The scattered light is detected by a photodiode, typically at 90° from the incident laser beam.

To understand the advantage of a universal detector such as the ELSD over the common UV–visible detectors, consider Figure 3.34. Note that because the saponins (compounds 7–10) do not absorb 254 nm electromagnetic radiation, they are not detected by the fixed-wavelength detector used in this study. In contrast, ELSD does not require absorbance but rather mere scattering, which all particles do.

It should be noted that the amount of scattered light depends on the scattering particle's size. Thus, ELSDs are much more sensitive to large molecules such as polymers and biomolecules. For this reason, they find more use coupled to SEC separations, which are also directed toward the separation of large molecules. ELSDs are seldom used in small-molecule applications. In these cases, photodiode array and mass spectrometry detectors are more common because of their simplicity and the additional analytical information they provide. Furthermore, ELSDs tend to suffer from small dynamic ranges due to the nonlinearity of the fundamental scattering processes upon which the technique is based.

 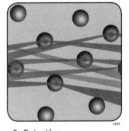

**1. Nebulization**
Column effluent passes through a needle and mixes with nitrogen gas to form a dispersion of droplets.

**2. Evaporation**
Droplets pass through a heated "drift tube" where the mobile phase evaporates, leaving a fine mist of dried sample particles in solvent vapor.

**3. Detection**
The sample particles pass through a cell and scatter light from a laser beam. The scattered light is detected, generating a signal.

(a)                              (b)                              (c)

**FIGURE 3.33** Schematic of an evaporative light scattering detector (ELSD). (a) Column effluent is nebulized after eluting from the column. (b) The aerosol droplets pass through a heated drift tube where solvent evaporation occurs. (c) Sample particle enter an optical cell where they pass through a laser beam. Scattering from the sample particles is detected and generates the electrical signal that generates the chromatogram. (*Source:* Reprinted from Alltech 800 ELSD Brochure #474, © 2003 with permission from Grace Davison Discovery Sciences.)

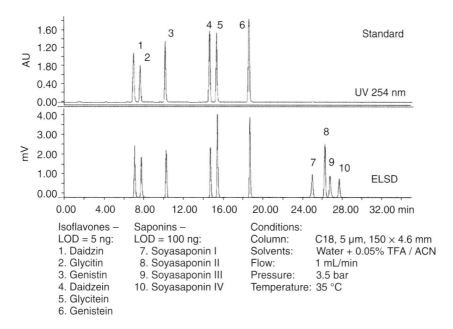

Isoflavones –        Saponins –                Conditions:
LOD = 5 ng:          LOD = 100 ng:             Column:        C18, 5 μm, 150 × 4.6 mm
1. Daidzin           7. Soyasaponin I          Solvents:      Water + 0.05% TFA / ACN
2. Glycitin          8. Soyasaponin II         Flow:          1 mL/min
3. Genistin          9. Soyasaponin III        Pressure:      3.5 bar
4. Daidzein          10. Soyasaponin IV        Temperature:   35 °C
5. Glycitein
6. Genistein

**FIGURE 3.34** Comparison of an evaporative light scattering detector (ELSD) to a UV detector at 254 nm used to analyze the same sample. Note that the saponins are not detected by the UV detector because they do not absorb 254 nm radiation, but they are detected by the ELSD, which is based simply on light scattering rather than on absorption. (*Source:* Reproduced with permission, Courtesy of Agilent Technologies, Inc. © Agilent Technologies, Inc. 2007.)

### 3.5.7.6. *Electrochemical Detectors.*

While many of the previous detectors rely on the interaction of analytes with electromagnetic radiation, amperometric detectors take advantage of the propensity of some molecules to be oxidized or reduced. Specifically, compounds such as phenols, mercaptans, peroxides, aromatic amines, ketones, aldehydes, nitro compounds, and conjugated nitriles are particularly susceptible to being oxidized or reduced. This redox chemistry is exploited by electrochemical detectors in which the mobile phase effluent is passed through an electrochemical cell containing working, auxiliary, and reference electrodes. An electrical potential is applied between the working and reference electrodes (see Figure 3.35). If the potential difference is such that it causes an analyte to be oxidized or reduced, an increased current flows in the cell as an analyte passes the working electrode. This change in current is measured and used to create a chromatogram based on current versus time. Because the current is the fundamental measurement, this type of detector is also commonly called an amperometric detector.

The advantage of electrochemical detection is that it is selective for species that can be oxidized or reduced, and this selectivity can be tuned even further by controlling the voltage applied to the electrodes such that it is above or below the redox potentials of the sample components. Due to their specificity, these detectors do not find widespread use. However, for certain applications they are quite valuable. For example, recent interest in brain chemistry requires information about neurotransmitter concentrations in the brain. Neurotransmitters are electrochemically active,

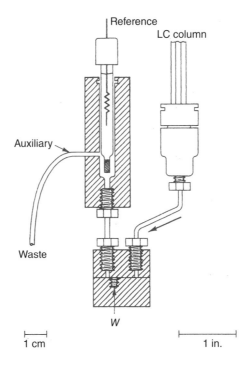

**FIGURE 3.35** Amperometric detector for LC. Solutes are oxidized or reduced at the working electrode (W), which creates a current that is measured to produce the chromatogram. (*Source:* Image courtesy of the copyright holder, Bioanalytical Systems, Inc.)

but are often contained in complex cell extracts. Thus, the selectivity of electrochemical detectors for electroactive species and their relative insensitivity to other sample components make electrochemical detectors attractive and effective detectors for such samples.

In addition to amperometric detection in which the analytes are chemically altered during the detection process, conductivity detection is another popular electrochemically based method. As discussed in the section on ion-exchange chromatography, solute ion zones can be detected by using ion suppression and monitoring the conductivity of the postsuppressor column effluent. This method does not rely on the oxidation or reduction of neutral analytes, but rather on the inherent electrical conductivity of ions.

***3.5.7.7. Detector Summary.*** Table 3.5 summarizes some of the key performance characteristics of common LC detectors.[51] Just as in GC, the demands of specific analyses must be considered when selecting an LC detector. As Table 3.5 suggests, there is no single detector that is appropriate for all applications. Rather, the detector must be chosen based on a knowledge of or an educated speculation about the solutes to be detected, their concentrations in the samples to be analyzed, and their response to the different modes of detection (i.e., Are they electroactive? Do they fluorescence? Can they absorb light?).

**TABLE 3.5  Characteristics of Common LC Detectors**

| Detector | Detection limit (g/mL) | Selectivity | Linear range |
|---|---|---|---|
| UV–visible absorption | $10^{-9}$ | General for solutes that absorb UV–visible radiation | $10^4$–$10^5$ |
| Fluorescence | $10^{-12}$ | Selective for fluorescent molecules | $10^3$–$10^4$ |
| Refractive index | $10^{-6}$ | Universal | $10^3$–$10^4$ |
| Mass spectrometry/SIM | $10^{-9}$/$10^{-12}$ | Universal or tunable to specific compounds in SIM | $10^4$–$10^5$ |
| Evaporative light scattering | $10^{-9}$ | Universal | $10^2$–$10^3$ |
| Amperometric | $10^{-10}$ | Selective for electroactive compounds | $10^4$–$10^5$ |

**EXAMPLE 3.3**

Select the detector that is best for detecting and quantifying each of the following. Explain your reasoning.

(a) Polyethyleneglycol (PEG) polymer chains
(b) Serotonin and dopamine in cell extracts
(c) Dyes in a mixture used to color clothing
(d) C-13 enriched benzene in a sample containing other related aromatic species
(e) Amino acids that can be covalently derivatized with fluorescent tags in a complex biological sample.

**Answers:**
(a) ELSD – good for detecting larger molecules and PEG is not highly absorbing in the UV–visible region.
(b) Electrochemical detection – they are both neurotransmitters and therefore electroactive.
(c) UV–visible – dyes by definition absorb well in the UV–visible region of the spectrum.
(d) Mass spectrometry – MS will differentiate between the masses of $^{13}$C and $^{12}$C.
(e) Fluorescence – derivatizing amino acids with fluorescent tags to make detection possible or more sensitive is a common technique.

## 3.6.  SPECIFIC USES OF AND ADVANCES IN LIQUID CHROMATOGRAPHY

In the following sections, we discuss some specific applications of liquid chromatography, focusing on two important topics: (1) separating chiral compounds and (2) preparative-scale chromatography. Both of these are vital to the pharmaceutical industry as well as other industries such as personal care products, food analysis, and natural supplements. We then discuss recent advances and trends in the field of liquid chromatography, including the development of high-speed separations and tandem liquid chromatography using two columns to achieve separations not possible on a single column. We end the chapter with a case study that illustrates how several of the key concepts presented in this chapter are put together to study the impact of pharmaceutical compounds released into the environment.

### 3.6.1. Chiral Separations

One of the major challenges facing analysts, particularly those in the pharmaceutical industry, is the analysis of chiral compounds. For example, consider the R and S enantiomers of thalidomide shown in Figure 3.36, which differ only in their geometric arrangement of side groups attached to the chiral center. From a general perspective, these two molecules are essentially identical in terms of their molecular weights, strengths of intermolecular interactions with bulk solvents, and physical characteristics (boiling points, electromagnetic spectra, mass spectra, etc.). Because the strength of the intermolecular interactions of the R and S enantiomers with a C-18 chain is essentially identical, ordinary RPLC cannot be used to separate and quantify these two compounds.

Thalidomide was prescribed in the late 1950s and early 1960s as a sedative. It was also given to pregnant women to treat morning sickness. However, many women using the drug gave birth to children with severe birth defects. From subsequent investigations, it is believed that the S enantiomer causes the birth defects and that the R enantiomer is the therapeutically active form. It should be noted that the R and S forms of thalidomide rapidly interconvert in the body, so administering an enantiomerically pure form of the drug does not eliminate the side effects.[52,53] Many drugs, however, do not interconvert *in vivo*. Thus, the ability to distinguish and separate drug enantiomers prior to their use in medications can be critically important. It should also be noted that thalidomide is still used to treat some medical conditions, including complications associated with leprosy. It has also been investigated for treatment of complications associated with cancer and HIV.[53,54]

This example, along with many others, illustrates why pharmaceutical companies are keenly interested in separating and quantifying the components of enantiomeric mixtures. In fact, because of examples such as thalidomide, the U.S. Food and Drug Administration

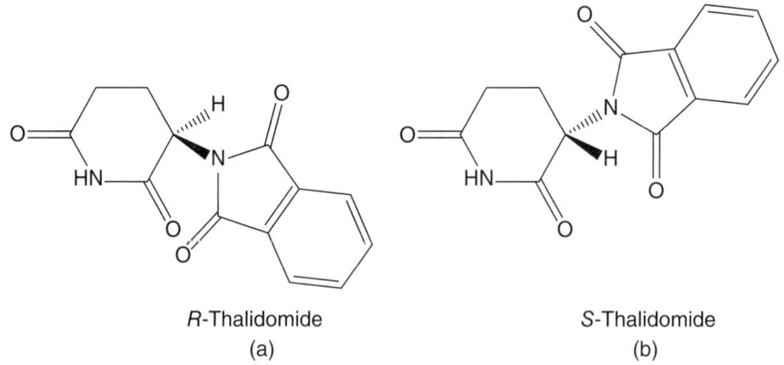

R-Thalidomide
(a)

S-Thalidomide
(b)

**FIGURE 3.36**   R & S enantiomers of thalidomide.

now requires chiral separations and analysis for all new pharmaceutical compounds. This requirement is made possible because our understanding of HPLC and chiral compounds has developed to the point where techniques exist for achieving chiral separations. To understand how chiral compounds are separated, it is important to recognize that chiral compounds have fixed orientations of side groups about a central chiral center. The key to chiral separations is to create stationary phases that can interact *differentially* with the enantiomeric pair of interest.

An analogy is appropriate here. Imagine immobilizing left-handed gloves on a stationary phase and passing a mixture of right and left hands through that stationary phase. Only the left hands have the proper orientation of fingers to fit into the immobilized gloves. Thus, the left hands, because of the stronger interactions with the stationary phase, are more retained than the right hands. If this separation were attempted on a nonchiral stationary phase, such as a C-18 column, both types of hands would be retained equally well because there would be no specific interactions with one hand over the other. However, on a chiral stationary phase, different selectivities exist and therefore a separation of enantiomers occurs.[55]

### 3.6.1.1. Chiral Stationary Phases (CSPs).

From a chemical perspective, consider attaching a chiral molecule to the stationary phase support, and then consider the interactions of two enantiomers with that stationary phase (Figure 3.37). Note that for one enantiomer, there is a B′–B interaction, whereas for the other enantiomer, a B–R interaction takes place. Suppose that the B and B′ groups are hydrogen bonding sites such as hydroxyl groups. The enantiomer that has the B′–B interaction will experience hydrogen bonding between the hydrogen atom on one molecule and the lone electron pairs on the oxygen atom of the other molecule. The other enantiomer will not experience this hydrogen bonding. The difference in the strength of attraction with the stationary phase results in a separation of the two enantiomers. Thus, the key to enantiomeric separations is to

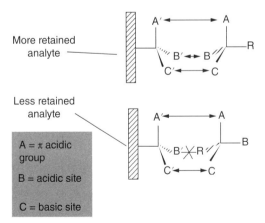

**FIGURE 3.37** Depiction of the different intermolecular interactions experienced by enantiomers on a chiral stationary phase. (*Source:* Reprinted from *Review of Chiral Separations* booklet with permission from Regis Technologies, Inc.)

introduce interactions that take advantage of the difference in orientations of the enantiomers. This usually requires three points of contact between the enantiomeric analytes and the chiral discriminating agent on the stationary phase. Many different types of chiral stationary phases (CSPs) exist. Some of them are listed in Table 3.6 and described in more detail as follows.

**Pirkle Phases.** The broadest class and most successful chiral separations were pioneered by Professor William Pirkle working at the University of Illinois.[56-60] Pirkle-type phases are based on small chiral molecules attached to silica support particles. An example of such a phase is naphthylleucine, shown in Figure 3.38. In this case, the aromatic ring structure acts as a $\pi$-electron donor that can specifically interact with $\pi$-electron-accepting side chains on enantiomers. Note that the backbone of this structure is the amino acid, leucine. All amino acids, except glycine, contain a chiral center and thus can exist as enantiomeric pairs. Thus, a number of phases using amino acids as the chiral selector are commercially available.

Figure 3.39 shows the chiral separations of some common medications such as ibuprofen, warfarin, and propranolol. Ibuprofen is an anti-inflammatory drug, warfarin (trade name Coumadin) is used as a blood thinner in patients who have suffered heart attacks, and propanolol is a widely used blood pressure regulator in a class of compounds known as $\beta$-blockers. All of these enantiomeric separations were achieved using Pirkle-type phases. Again, the importance of these separations is that our bodies can respond differently to the two different enantiomeric forms of a drug, as was the case with thalidomide, so being able to identify the presence of isomers and to isolate one form from the other is incredibly important.

**TABLE 3.6  Some General Classes of Chiral Stationary Phases**

| Chiral phase class | Chiral selector |
| --- | --- |
| Pirkle phases | Small chiral molecules |
| Cyclodextrin phases | Cup-shaped cyclic glucose structures |
| Polysaccharide phases | Derivatized amylose and cellulose |
| Macromolecular phases | Chiral macrocyclic molecules and proteins |

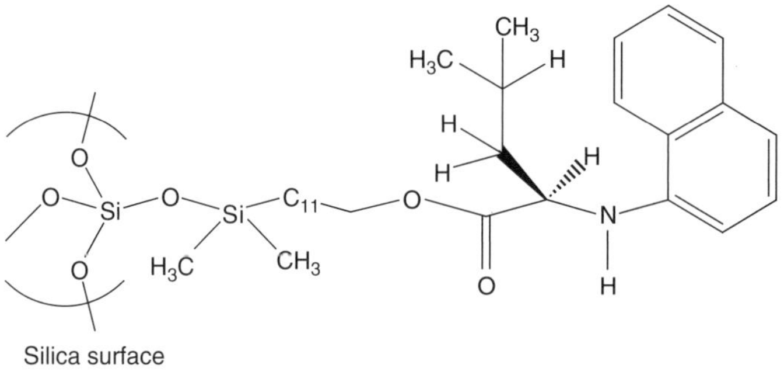

Silica surface

**FIGURE 3.38**  An example of a Pirkle-type chiral stationary phase; L-Naphthylleucine. (*Source:* Reproduced with permission of Regis Technologies, Inc.)

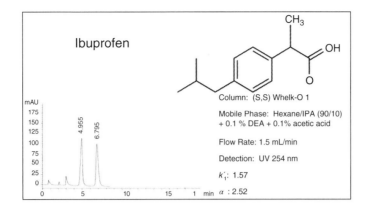

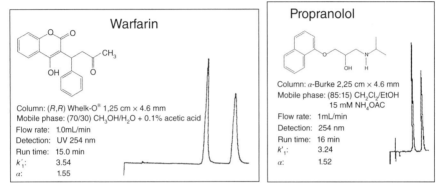

**FIGURE 3.39** Examples of separations of enantiomers using Pirkle-type chiral stationary phases. (*Source:* Reproduced with permission from Regis Technologies, Inc.)

**Polysaccharide Phases.** Another set of highly successful CSPs is based on derivatized polysaccharide-coated silica gel particles and is sold under the Chiralcel and Chiralpak trademarks. These phases use derivatives of amylose (polymers of α-1,4 linked glucose) or cellulose (polymers of β-1,4 linked glucose). The chiral selectivity is brought about in part by the ability of the phase to form hydrogen bonds with analytes and by the helical conformation of the polysaccharide backbone that can preferentially inhibit access of one enantiomer over the other to hydrogen-bonding sites. Changing the mobile phase composition alters the three-dimensional structure of the polysaccharide and thus provides a means of controlling selectivity. Because of the success achieved with them, these phases are the most commonly used CSPs for chiral separations.

**Cyclodextrin Phases.** Other successful separation methods have been developed in addition to the Pirkle-type phases that historically paved the way for chiral separations. Some of these phases use derivatized amino acids covalently linked to support particles to provide the chiral discrimination.[61–63] Others are cyclodextrin-based stationary phases. Common cyclodextrins are shown in Figure 3.40. Cyclodextrins are cup-shaped molecules comprised of six, seven, or eight glucopyranose rings. The internal cavity is relatively hydrophobic, while the exterior is relatively hydrophilic due to the presence of —OH groups. The different forms of enantiomeric pairs can orient differently within the cavity

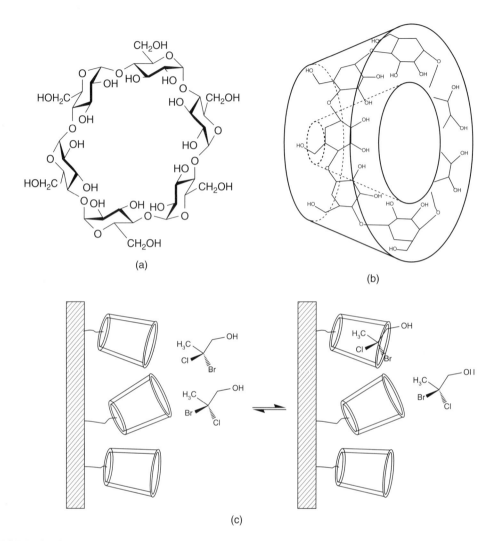

**FIGURE 3.40**   (a) $\alpha$-Cyclodextrin ($\beta$- and $\gamma$-cyclodextrin have 7 and 8 glucose units). (b) Depiction of the cup-like 3D shape of cyclodextrins. (c) Depiction of cyclodextrins chemically bonded to the surface of a chromatographic support particle and their ability to selectively bind one enantiomer more strongly than the other. (*Source*: Reprinted with permission of http://www.chem.qmul.ac.uk/iupac/2carb/37 .html (a))

and therefore experience different blends of intermolecular interactions, ultimately resulting in a separation of the enantiomeric pair.

   **Macrocyclic and Protein Phases.** Macrocyclic antibiotics such as vancomycin (see Figure 3.41) and teicoplanin have also been covalently linked to silica particles to create CSPs.[64–66] These phases allowed for separations that had not been achieved on other phases and further extended the range of options for separating chiral compounds via HPLC. An example of the separations achieved is shown in Figure 3.42. It is important to keep in mind that the compounds being separated differ only in the orientation of groups around chiral centers. Thus, these separations are really quite remarkable and impossible to achieve on C-18 columns.

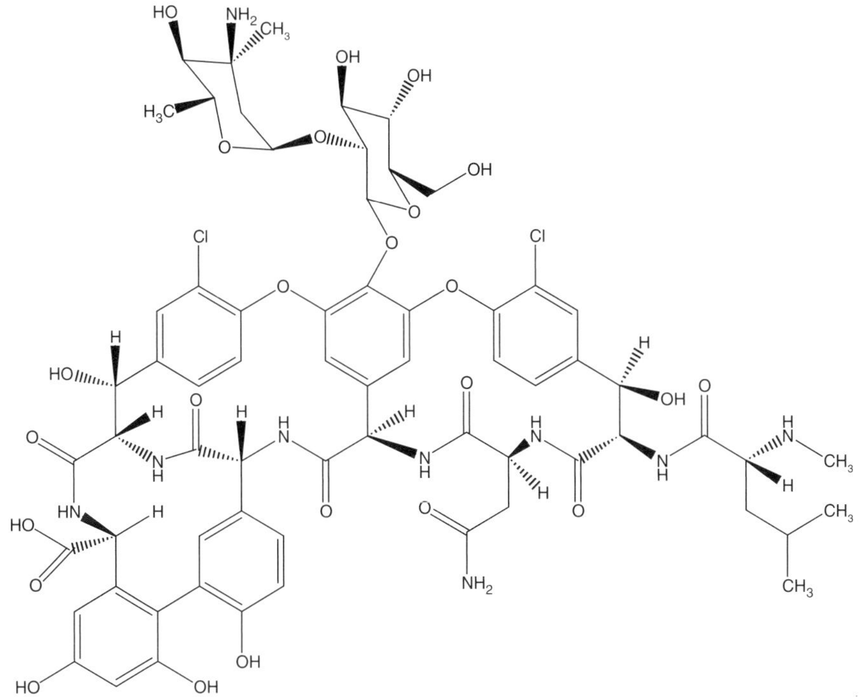

**FIGURE 3.41** Structure of vancomycin. Note the number of chiral centers and the multiple sites capable of donating and accepting hydrogen bonds. Because of these numerous chiral and hydrogen bonding sites, the molecule is able to bind different enantiomers of the same molecule with different strengths and thus help achieve their separation.

Still other CSPs are based on proteins that are covalently bonded to silica supports. The proteins used include α-chymotrypsin, ovomucoid, pepsin, cellobiohydrolase, and human serum albumin (HSA). These proteins offer chiral recognition sites and thus provide the selectivity required to separate enantiomeric pairs.

***3.6.1.2. Chiral Mobile Phases.*** While the use of chiral stationary phases is the most common method of achieving chiral separations, it is also possible to modify the mobile phase instead. In these cases, chiral compounds or cyclodextrins are added to the mobile phase, where they can interact with the chiral analytes. Again because of differential binding of the enantiomeric forms to the organic modifier, the enantiomers' affinities for the mobile phase differ. The enantiomer that interacts more favorably with the mobile phase modifier is drawn out of the stationary phase more easily and thus spends less time being retained than the weaker-interacting enantiomer. Thus, the addition of chiral selectors to the mobile phase can also be used to achieve chiral separations.

## 3.6.2. Preparative-Scale Chromatography

The discussion up to this point has focused on what is known as analytical-scale chromatography in which small volumes (microliters) of relatively dilute (millimolar to micromolar) analytes are analyzed. Information about solute identity and concentration can be obtained,

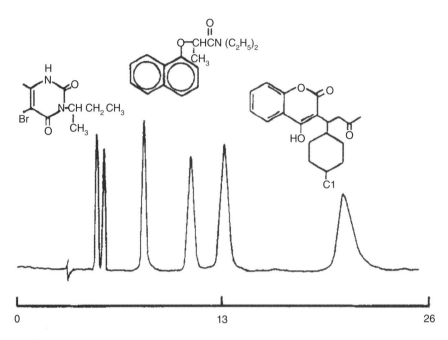

**FIGURE 3.42** Reversed-phase enantiomeric separation of (from left to right) bromacil, devrinol, and coumachlor on a vancomycin chiral stationary phase. Column: $25\,cm \times 0.44\,cm$ column, Silica particles: 5-μm, mobile phase: 10:90 acetonitrile: 1% pH 7 triethylammonium acetate buffer (v/v). Flow: 1.0 mL/min, temp: ambient, detection: UV at 254 nm. (*Source:* Reprinted in part with permission from *Anal. Chem.*, 1994, 66, 1473–1484. Copyright 1994 American Chemical Society.)

but the solute molecules themselves are not collected or isolated; with such small volumes and low concentrations, there is simply not enough to collect, isolate, and subsequently use.

Preparative-scale chromatography, however, aims at the purification and collection of milligram to gram quantities of material. While the goal of analytical chromatography is qualitative and quantitative analysis, the goal of preparative chromatography is the collection of the maximum amount of pure substance per injection.[67]

Preparative-scale chromatography is conducted using stationary phases that are similar to those used in NPLC and RPLC analytical separations, with C-18 stationary phases being particularly popular in preparative-scale work. Several important operational differences exist, however, between analytical and preparative-scale separations. Specifically, in analytical separations the amount (mass) of analyte injected is intentionally kept low so as to remain in a region where Langmuir-like isotherms are obeyed, or more generally, that the solutes act as if they are at infinite dilution in the mobile and stationary phases. This means that the solute molecules do not interact with one another and therefore do not influence each other's retention behavior. If this condition is exceeded and large amounts of solute are injected, the stationary phase becomes flooded with analyte molecules and the column is said to be "overloaded." The stationary phase then does not have enough capacity to retain all of the analyte molecules of the same structure equally and the retention behavior changes. In addition, peak shapes become distorted due to the nonlinear isotherm behavior.

In preparative-scale chromatography, the column is intentionally overloaded so as to maximize the amount of sample loaded onto the column and thereby maximize the yield of the separation. Furthermore, much larger columns are used in preparative scale work than in analytical separations. Columns with diameters from 4.6 to 50 mm and lengths from 50 to 250 mm are often used. The larger the diameter, the more sample can be loaded onto the column. However, large columns cost hundreds to thousands of dollars and require more solvent.

The particles used for preparative scale work are similar to those for NPLC and RPLC in terms of their chemical properties and modes of retention. However, to decrease the pressure required to pump mobile phase through the column, larger particles (5–10 μm) are used in preparative scale applications compared to the desirable small particles (1–5 μm) found in analytical separations.

Because preparative scale columns are expensive and use considerable amounts of solvents, separations using them are often first approximated by deliberately overloading analytical scale columns to develop and optimize the separation, followed by extending the methodology (i.e., mobile phase and stationary phase conditions) to the preparative scale column.

Figure 3.43 shows the development of a preparative scale isolation of caffeine from theophylline.[67] Note the nearly Gaussian peak shapes obtained with small injection amounts on the analytical column, and note how the front edges of the peaks distort as the column becomes overloaded. The flat tops of the peaks arise because virtually all of the light passing through the detector cell is being absorbed by the sample, causing a plateau in the absorbance readings. In part "b" of Figure 3.43, it is clear that caffeine and theophylline are well separated on the preparative scale column.

To isolate the compounds of interest, the column effluent is collected using separate containers to collect different fractions of the effluent. In Figure 3.43, the column effluent from about 1.8 to 4.2 min would be collected in one vial, while a different vial would be used to isolate the theophylline eluting between 6 and 8 min. Of course, the effluent still contains the mobile phase solvent (in this case, water and acetonitrile), so to isolate the pure components the mobile phase solvents must be removed, typically by volatilization, leaving the less volatile pure analyte behind.

Because solvent removal is required, NPLC-based preparative scale separations are advantageous because they use mobile phases comprised of volatile organic solvents that are easily vaporized. However, because of the ubiquity and familiarity of RPLC analytical separation, reversed-phase preparative scale separations are still the most common. Other modes of preparative scale work include ion-exchange, size exclusion, and chiral separations, but these typically employ specialty columns and can be considerably expensive. Nevertheless, some applications, especially protein and polymer purification, may require their use.

One final advantage regarding the intentional overloading of columns should be noted. Figure 3.44 shows the separation of three antibiotics.[67] Note that when small masses are injected, only three peaks are detectable, leading one to believe that the antibiotic samples are pure. However, as more sample is injected, early-eluting impurities become increasingly easier to detect (note also the degradation of peak shapes as the column gets overloaded). These impurities are present in the antibiotics but simply were not

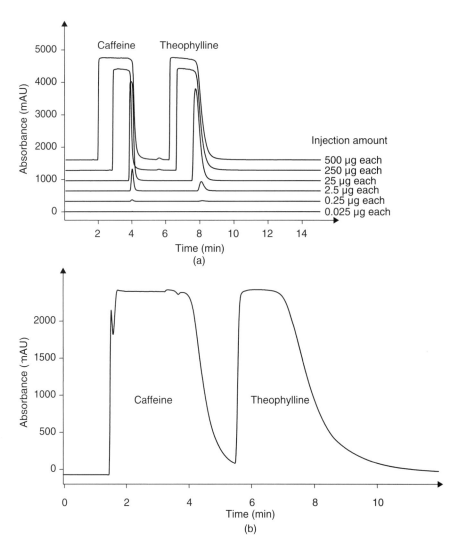

**FIGURE 3.43** Scale up of a reversed-phase separation of two xanthines. Shown are (a) analytical column separations and (b) the preparative separation resulting from scale-up calculations. (a) Column: 150 mm × 3 mm, 5 μm $d_p$ Zorbax SB-C18 (Agilent Technologies, Wilmington, Delaware); mobile phase: 90:10 (v/v) water-acetonitrile; flow rate: 0.6 mL/min; detection; UV absorbance at 270 nm; pathlength: 10 mm; ambient temperature. (b) Column: 150 × 21.2 mm, 5 μm $d_p$, Zorbax SB-C18; mobile phase: 90:10 (v/v) water-acetonitrile; flow rate: 25 mL/min; detection; UV absorbance at 270 nm; pathlength: 3 mm; ambient temperature. (*Source:* Excerpted and reprinted with permission from *LC/GC North Am.*, 2004 (22) 422. LC/GC North America is a copyrighted publication of Advanstar Communications Inc. All rights reserved.)

detected when small amounts of the antibiotics were injected. The identification and removal of such impurities is critically important to pharmaceutical companies because such impurities can have adverse health effects. It is therefore important to conduct deliberate overloading studies to search for impurities. Once identified, preparative scale chromatography can be used to eliminate these impurities from the final isolated pure products.

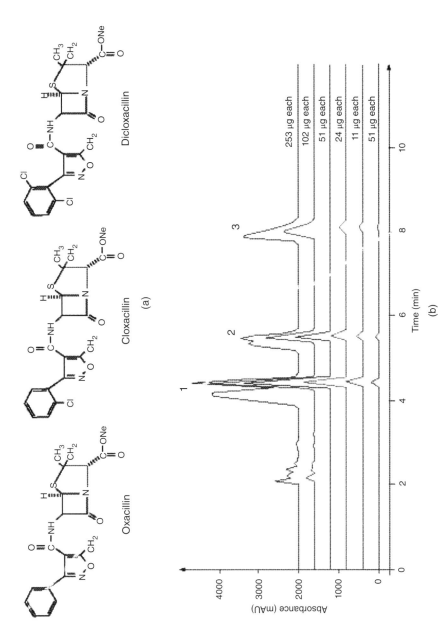

**FIGURE 3.44** Scale-up of a separation of three antibiotics. (a) The structures of the antibiotics and (b) scale-up separations on an analytical column. (a) Column: 150 mm × 4.6 mm, 5 μm $d_p$ Zorbax Eclipse XDB-C18; mobile phase: 65:35 (v/v) water–acetonitrile, both with 0.1% trifluoroacetic acid; flow rate: 1 mL/min; detection; UV absorbance at 254 nm; injection volume 30 μL; ambient temperature. (*Source:* Excerpted and reprinted with permission from *LC/GC North Am.*, 2004 (22) 422. LC/GC North America is a copyrighted publication of Advanstar Communications Inc. All rights reserved.)

### 3.6.3. Ultra-High Performance Liquid Chromatography (UHPLC) for High-Speed Separations

Time is money, and so is paying for chemical waste disposal. These two forces have driven chromatographers to reduce HPLC analysis times and the amount of waste generated. This has led to the use of shorter (2–5 cm) and narrower columns (2–3 mm) as opposed to the standard 15–30-cm-long, 4.6-mm-wide columns. Reducing the length of the column results in a shorter distance to travel for the analyte and therefore requires less time and mobile phase to elute the solutes. Narrower columns also reduce the volume of mobile phase used.

While decreasing the column length makes sense in terms of having shorter retention times, the practical problem is that the number of theoretical plates (N) is proportional to column length. Thus, dramatically decreasing the column length also dramatically reduces the resolving power of the column. While solutes elute faster on short columns, they may also overlap or coelute, ruining the analysis and eliminating any potential cost savings.

---

**EXAMPLE 3.4**

Calculate the resolution of two peaks on a 25.0 cm column with 10,500 plates for two solutes with retention factors of 3.75 and 4.15. Assuming nearly identical peak heights, will the two compounds be baseline resolved under these conditions?

**Answer:**

$$R = \frac{\sqrt{N}}{4} \left( \frac{\alpha - 1}{\alpha} \right) \left( \frac{k}{1 + k} \right)$$

$$\alpha = \frac{k_B}{k_A} = \frac{4.15}{3.75} = 1.107$$

$$R = \frac{\sqrt{10,500}}{4} \left( \frac{1.107 - 1}{1.107} \right) \left( \frac{4.15}{1 + 4.15} \right) = 2.00$$

A resolution of 1.5 represents baseline separation, so the two compounds would be baseline resolved.

Repeat the calculation assuming everything stays the same except that, in an effort to decrease the analysis time, the column is reduced from 25.0 to 5.0 cm. Will the compounds still be resolved?

**Answer:**

Because the number of plates is proportional to the length of the column, decreasing the length to one-fifth of the original decreases $N$ by the same amount. So now, $N = 2100$.

$$R = \frac{\sqrt{2100}}{4} \left( \frac{1.107 - 1}{1.107} \right) \left( \frac{4.15}{1 + 4.15} \right) = 0.89$$

No, the peaks will no longer be baseline resolved. They will overlap to some degree.

Ultra-high performance liquid chromatography (UHPLC) uses several techniques to overcome this limitation. The technology now exists to manufacture smaller packing particles reproducibly. Smaller particles yield considerably less band broadening, especially when used at high mobile phase velocities.[68] Specifically, the resistance to mass transfer in the mobile phase term in the plate height equation (see Chapter 1, page 35) depends on the *square* of the particle diameter. Thus, even modest decreases in the particle diameter have dramatic effects on $N$ and therefore on resolution, $R$ (see the general resolution equation in Chapter 1, page 47). The effect of particle size on the plate height of a column is illustrated in Figure 3.45.[68] Smaller particle diameters offset some of the loss of theoretical plates incurred by using shorter columns.

Another important practical consequence of Figure 3.45 involves the higher mobile phase velocities that can be used with small particles. It is clear that as the velocity (related to flow rate in the figure) is increased when using large particles, the HETP increases dramatically. Because HETP $= L/N$, as HETP increases, $N$ decreases and so does resolution. There is very little increase in HETP as the velocity (flow rate) is increased with particles that are smaller than 3 μm in diameter, as well as with the superficially porous particles described above. Thus, there is little decrease in $N$ and almost no change in resolution upon changing the flow rate from 1 to 6 mL/min. Because analysis time is nearly proportional to flow rate, operating at 6 mL/min compared to 1 mL/min reduces the analysis time nearly sixfold. *The smaller particles therefore allow for the use of shorter columns at high mobile phase flow rates with little sacrifice in resolution.*

How fast can fast be? Consider Figure 3.46 that shows a three-component separation on a 15 cm column with 5 μm particles, compared to a fast separation on a short 3 cm column with 3 μm particles.[68] There is nearly a 10-fold decrease in analysis time with very little loss of resolution. This means that 10 analyses of such samples could be achieved on the short column in the time it takes to do a single analysis on the standard column (ignoring the

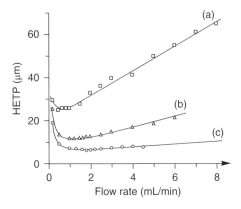

**FIGURE 3.45** Effect of particle size on plate height (i.e., the height equivalent to a theoretical plate, HETP). (a) A 250 mm × 4.6 mm column packed with 10-μm particles, (b) a 125 mm × 4.6 mm column packed with 5 μm particles, and (c) a 100 mm × 4.6 mm column packed with 3 μm particles. (*Source:* Excerpted and reprinted with permission from *LC/GC North Am.*, 2002 (20) 40. LC/GC North America is a copyrighted publication of Advanstar Communications Inc. All rights reserved.)

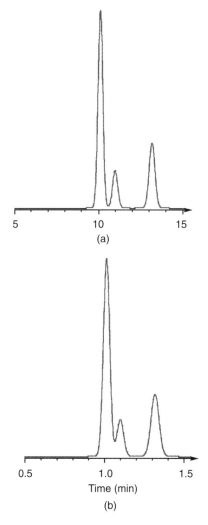

**FIGURE 3.46** Simulation of a three-component separation obtained using (a) conventional and (b) fast LC conditions. Note the very different time scales in each chromatogram (15 min vs 1.5 min). Column (a): 150 mm × 4.6 mm, 5 µm particles, flow rate: 1.0 mL/min; column efficiency: 5400; operating pressure: 1200 psi; minimum resolution: 1.51; extracolumn volume: 70 µL. Column (b): 30 mm × 4.6 mm, 3 µm particles, flow rate: 2.0 mL/min; column efficiency: 2700; operating pressure: 1900 psi; minimum resolution: 1.07; extracolumn volume: 7 µL. (*Source:* Excerpted and reprinted with permission from *LC/GC North Am.*, 2002 (20) 40. LC/GC North America is a copyrighted publication of Advanstar Communications Inc. All rights reserved.)

time it takes to load and inject the sample, which can be considerable as discussed below). For this reason, fast HPLC is also referred to as "high-throughput" chromatography.

Fast separations require special modifications to standard HPLC systems in order to preserve the resolution obtained in a short column and to reduce the degradation resulting from broadening effects that occur in instrument components outside of the column (known as extra-column broadening).[68] In conventional HPLC systems, the majority of band broadening occurs within the column, so extra-column broadening effects are seldom of great concern. This is not the case when one uses a conventional system with short columns for fast analyses. The columns retain solutes for such short amounts of time that *broadening contributions from the transfer lines (e.g., the tubing between the injector and the column, and between the column and the detector), the injector, and the detector become comparable to those from the column and can significantly degrade the overall resolution.*[68] This extra-column broadening arises from the parabolic flow profile that exists in open tubes, such as the transfer lines. As was discussed in Chapter 1, parabolic flow spreads out the solutes along the column, with increased velocities leading to increased broadening. So before the peaks reach the column, they are already being broadened within the transfer lines. The peaks also broaden while in the column due to diffusion and eddy dispersion effects. The total broadening is the sum of all of the contributions within the individual parts of the LC system. As the contribution to broadening from the column decreases, the broadening contributions from the other components become relatively more significant.

In order to address this problem, the length and diameter of the transfer lines are kept to a minimum in UHPLC systems. Incorporation of microflow UV detector cells that reduce the detector volume from 8 μL to submicroliter volumes decreases band broadening in the detector.

If a mass spectrometric detector is used with high-speed LC, common LC–MS interfaces cannot handle the high mobile phase flow rates because solvent is being delivered faster than it can be vaporized. This is problematic in terms of maintaining the low pressure required for proper mass spectrometer performance. One solution to this is to split the column effluent flow into two streams, one directed into the LC–MS interface and the other to waste. This, however, results in loss of analyte molecules, which can limit the detection of low-concentration analytes. One way around splitting the flow is to use narrow bore columns (1, 2, and 3 mm diameters) instead of the conventional 4.6 mm columns. Achieving the same linear velocity in microbore columns as in the 4.6 mm columns allows for much lower volumetric flow rates without a loss in column efficiency or resolution. For example, a 4.6-mm-diameter column operated at 2 mL/min can be replaced with a 2-mm-diameter column operated at 400 μL/min to achieve comparable separations, reducing the volume of solvent delivered to the MS interface to a manageable level. This, however, requires a pumping system that can reliably deliver fractions of a milliliter per minute. Conventional pumps cannot usually maintain constant flow rates at such low velocities.

Another reason that conventional pumps cannot be used is the high back pressures created by the use of smaller particles. Smaller particles pack together such that the spaces between particles are smaller than with large particles. These smaller through-pores make it more difficult to pump liquid through them (imagine trying to blow water through a straw packed with fine sand versus larger pebbles). Therefore, pumps must be capable of reaching higher pressures in order to push mobile phase through columns packed with small particles. Instruments designed for UHPLC are manufactured with such pumps.

Finally, the *total* time it takes to analyze a sample includes not only the time the sample is on the column but also the time it takes to position the sample for injection, load the sample into the injector, inject the sample, and wash the sample needle in preparation for the next analysis. If all of these steps take considerably longer than the actual separation time, the benefits of short chromatographic times are greatly diminished. As a consequence, autoinjector systems for fast HPLC systems are modified relative to their conventional counterparts, which typically take between 1 and 4 min to perform an injection. Fast autosampling systems are equipped with faster robotic components that take less time to select, position, and inject the sample.[68] The number of vials an autosampler tray contains can also limit the utility of high-speed LC. For example, if an entire run can be accomplished in 4 min, then a 12-h overnight automated sequence could analyze 180 different samples (assuming no replicates). Thus, the capacity of the sample tray must accommodate the increased number of analyses per unit time.

The above paragraphs implicitly suggest that choosing the optimum column length, column diameter, particle size, and mobile phase velocity (as it relates to pump pressure) are interrelated decisions. This is true and much work has gone into predicting how to get the maximum performance out of HPLC systems.[69,70] These considerations are beyond of the scope of this text, but the references show how all of these parameters can be simultaneously optimized to achieve the highest number of theoretical plates possible in a given period of time, or achieve a desired number of theoretical plates in the shortest time possible.

### 3.6.4. Tandem-Column Liquid Chromatography

The general resolution equation shows that very small increases in the separation factor between two solutes can result in dramatic improvements in resolution. In most applications, changes in mobile phase composition can be used to adjust the separation factor; however, sometimes mobile phase changes alone cannot provide the required separation. The use of two chemically distinct stationary phase columns connected in series can provide increased control of the separation factor, comparable to the approach taken in tandem-column GC systems.[71, 72]

In tandem-column GC, the pressure drop across the two different columns is used to control the relative influence of each column on the overall separation. In tandem-column LC, temperature can be used to control the relative influences.[71] A block diagram of a tandem-column LC system is shown in Figure 3.47.[71]

The key to using tandem-column LC is that the two columns must offer different intermolecular interactions between the solutes and the stationary phases.[72] For example, it would be of little value to attach a C-8 column to a C-18 column because their interactions with solutes are essentially the same. A better combination would be a C-18 column followed by one filled with particles coated with a thin layer of carbon atoms, for example, carbon-coated zirconia. Solute retention on C-18 is driven by a partition mechanism while

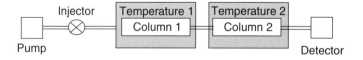

**FIGURE 3.47**  Block diagram of a tandem-column LC system. The contribution of the two columns to the overall solute retention is controlled by controlling the temperature of each column. Lower temperatures lead to greater contributions to retention. (*Source:* Excerpted and reprinted with permission from *LC/GC North Am.*, 2003, 21, p. 158 and from the authors. LC/GC North America is a copyrighted publication of Advanstar Communications Inc. All rights reserved.)

retention on carbon phases is driven by adsorption to the surface. Thus, the two retention mechanisms are distinct and provide different interactions for different components.

The selectivity of dual-column systems is further controlled by the temperatures of the two columns. Higher temperatures lead to decreased retention. Thus, increasing the temperature of one column relative to another reduces the contribution of that column to the overall retention of solutes. In this way, controlling the relative temperatures of the two columns allows for the precise adjustment of the overall separating ability of the column ensemble and makes it possible to achieve separations not achievable with the limited selectivities provided by a single column.

Figure 3.48 demonstrates the improved resolving power of tandem-column methods compared to single columns.[71] The separation of 10 herbicides is accomplished in 10 min

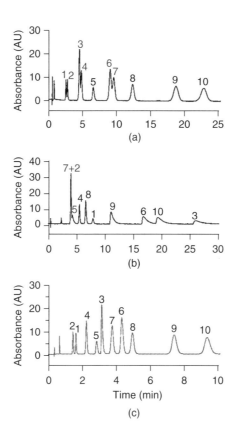

**FIGURE 3.48** Chromatograms showing the separation of a mixture of triazine herbicides on (a) C-18 at 30 °C and a 1-mL/min flow rate, (b) carbon-coated zirconia at 60 °C and a 1 mL/min flow rate, and (c) a thermally tuned tandem set with a C-18 column at 30 °C and a carbon-coated zirconia column at 125 °C, and a 3 mL/min flow rate. Mobile phase: 30:70 (v/v) acetonitrile–water; detection wavelength: 254 nm. Peaks: 1 = simazine, 2 = cyanazine, 3 = simetryn, 4 = atrazine, 5 = prometon, 6 = ametryn, 7 = propazine, 8 = turbulazine, 9 = prometry, 10 = terbutryn. (*Source:* Excerpted and reprinted with permission from *LC/GC North Am.*, 2003, 21, p. 158 and from the authors. LC/GC North America is a copyrighted publication of Advanstar Communications Inc. All rights reserved.)

in a tandem mode, whereas the same compounds cannot be resolved in 30 min on either of the individual columns.

One drawback to tandem-column LC is that in order to have maximum flexibility, at least one of the columns must be heated to elevated temperatures (60–250 °C) in order to achieve a significant blend of column influences that lead to the desired separations. Unfortunately, silica particles tend to dissolve in aqueous mobile phases when operated for long periods at elevated temperatures. In this regard, zirconia-based particles offer much greater stability at temperatures up to 200 °C for prolonged times. Thus, it is possible to couple a silica-based column operated at relatively low temperatures (below 60 °C) to a zirconia-based column operated at elevated temperatures (60–200 °C) and have reproducible, stable chromatographic systems.[71]

### 3.6.5.  Two-Dimensional Liquid Chromatography (2D-LC)

It is often impossible to find a single set of conditions that completely separates all of the components in complex mixtures such as petroleum products, food extracts, and biological samples. Peaks frequently overlap in such situations, either significantly reducing resolution or entirely obscuring the fact that some peaks are due to multiple, coeluting components.

In such cases, two-dimensional liquid chromatography (2D-LC) is beneficial. In 2D-LC, small portions of the effluent from one column (the first dimension) are transferred to a second column for further separation (the second dimension). It is easiest to start by imagining the process being done manually, or what is known as the "offline" mode. Suppose that the separation shown in Figure 3.49 is obtained on a single column. Further suppose that as the mobile phase and solutes elute from the column, they are collected in a vial. After each 30-s period, the vial is removed and replaced with a new vial. Now, each of the 12 separate vials that were collected over the 6-min first-dimension analysis can be analyzed on a second column under different conditions (e.g., different stationary phase, mobile phase composition, etc.). Vials that looked like they contained only a single compound because only a single peak appeared may actually be found to contain multiple components.

The key to 2D-LC is that the second separation needs to be based on different intermolecular interactions than the first. If the second column is identical to the first, all of the components that coelute on the first will just coelute again on the second. In chromatography, the term "orthogonal" is used to describe two mechanisms of separation that are based on substantially different intermolecular interactions.[73] HILIC and RPLC are an example of a pair of mechanisms that are highly orthogonal. In RPLC, hydrophilic compounds tend to elute early, whereas in HILIC, hydrophilic compounds are more retained.

Figure 3.50 shows the results from an offline HILIC × RPLC separation of an extract of a traditional Chinese medicine.[74] Spending time understanding this figure will prove valuable to understanding 2D-LC. For example, imagine collapsing or compressing all of the peaks from right to left, such that all of the resulting peaks are shown just on the left axes (the one labeled "fraction"). Such an image would be the chromatogram collected if just the first dimension was used to analyze the sample. Clearly, some of the peaks would

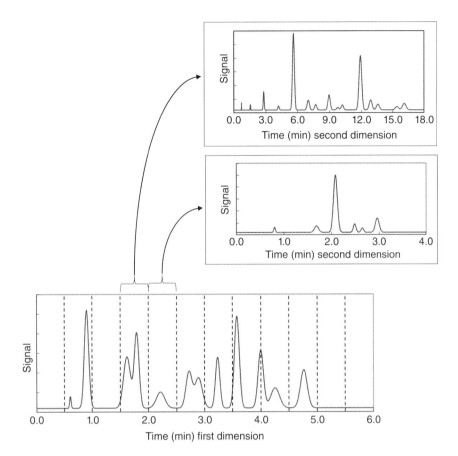

**FIGURE 3.49** Simulation of comprehensive sampling of the effluent from one column (the first dimension) with subsequent analysis of that portion on a second column (the second dimension). In this case, samples are collected every 30 s, as indicated by the dashed lines. The interest in 2D-LC lies in the potential to observe compounds that partially or entirely overlapped under one set of chromatographic conditions by further separating them under a substantially different set of conditions. Here, the subsequent analysis of only two portions of the first dimension is being shown, but all eight portions that clearly contain at least one compound could be analyzed with the second column for comprehensive analysis.

be on top of one another, meaning that what appeared to be single peaks were actually caused by compounds that had not been separated. Another thing to do with this figure is to follow any of the individual fractions from left to right. For example, look at Fraction 7. As you move from left to right along the Fraction 7 line (i.e., up the time axis), two prominent peaks are observed with several other smaller peaks – for a total of about 15–20 different molecules that eluted while Fraction 7 was collected. Most of the other fractions also show multiple compounds that coeluted or were not fully resolved by the first dimension. So signals that appear to be a single peak in one dimension may actually contain multiple components! Lastly, we note that the vertical mAU scale (milli absorbance

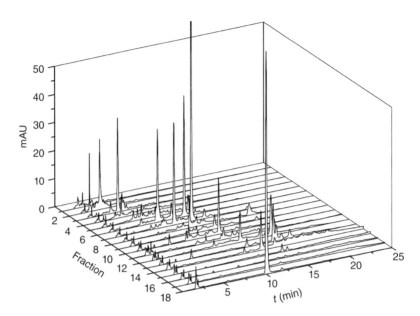

**FIGURE 3.50** Offline HILIC × RP separation of a Chinese medicine extract. Fractions from the first dimension were collected once every minute and subsequently analyzed by RPLC on a C-18 column. Each line running from left to right therefore represents a separate RPLC separation. If the peaks are compressed or condensed onto the left axis labeled "Fraction" by mentally sweeping them from right to left, one would have an idea of what the chromatogram from the first dimension looked like. Clearly, by looking at some fractions (like 7 and 10), many compounds eluted during the minute that the effluent was collected. Thus, it is likely several peaks partially or entirely overlapped, obscuring the true complexity and chemical makeup of the sample. The third axis (mAU) is the detector response at 280 nm. (*Source:* Reprinted from Z. Liang, et al. *J. Chromatogr. A*, 2012, 1224, 61–69 with permission from Elsevier.)

units) is the magnitude of the response of a UV detector operating at 280 nm and provides quantitative information if standards with known concentrations are analyzed on the same system.

***3.6.5.1. Peak Capacity.*** To understand the advantage of 2D-LC over 1D-LC, it is necessary to introduce the concept of peak capacity, symbolized by $n_c$. The peak capacity of a single column (i.e., 1D-LC) is simply the maximum number of peaks that can be resolved over a given period of time. In 2D-LC separations, the peak capacity is given by

$$n_{c,2D} = {}^1n_c \times {}^2n_c \tag{3.9}$$

where the superscripts indicate the two different columns or dimensions. This equation makes it clear that 2D separations have the potential to separate a far greater number of components compared to either individual column alone. This arises because each portion from the first column is analyzed with the second so that any peaks that overlapped in the first dimension can be separated by the second.

Returning to the discussion above regarding offline 2D-LC, suppose that the first column has the capacity to separate 25 individual peaks. The effluent collected as each of these 25 peaks eluted could then be analyzed on the second column. If the second column can then separate, say, 15 components (assuming that the single peaks from the first dimension were actually due to mixtures of coeluting species), then the total peak capacity of the entire separation is $25 \times 15 = 375$ peaks – which is substantially greater than either column on its own. This is the power of 2D-LC.

### 3.6.5.2. Online 2D-LC.
In practice, manual, off-line 2D-LC is tedious, labor intensive, and time consuming. It is much more practical to perform the operation "online" using instrumentation designed to collect fractions eluting from the first column and subsequently inject them onto the second column. This is typically achieved using multiport valves. These function such that a specified volume of effluent from the first column is captured by a sample loop. At the appropriate time, a valve rotates, which does two things simultaneously. First, the sample from the first column that had just been collected is introduced onto the second column. Second, a fresh sample loop is put in place to collect sample from the first column while the previous sample is being analyzed on the second. This process repeats throughout the entire separation. This requires that the separation in the second dimension be completed in the time that it takes to collect the sample from the first dimension. So the second dimension separation must be fast, which is partly what drove interest in the development of high-speed separations discussed above. Examples of the results of online 2D-LC analyses are shown in Figure 3.51.[75,76]

### 3.6.5.3. Comprehensive, Heartcutting, and Selective Comprehensive Sampling.
The method of continuously sampling the first-dimension effluent and passing it onto the second dimension as described above is known as "fully comprehensive" sampling and is symbolized by LC×LC (note the times sign). Another mode of sampling is also common and is known as "heartcutting." In this mode, only a single or very small number of specific regions (time spans) of the first-dimension chromatogram are collected and analyzed on the second column. This allows a particular peak from the first dimension to be targeted for subsequent analysis. This mode is given the symbol LC–LC (note the dash instead of the times sign). A third mode, called "selective comprehensive" LC×LC (sLC×LC) blends the two approaches by taking multiple samples during the elution of particular peaks. These three methods are depicted in Figure 3.52. Each mode has its advantages and disadvantages, which are discussed in more depth by Groskreutz and Stoll.[77]

### 3.6.5.4. HILIC and Mass Spectrometry with 2D-LC.
It is worth noting that using HILIC as the second dimension in a 2D separation not only provides a separation mechanism that is orthogonal to RPLC, but it also makes detection by mass spectrometry highly attractive. Recall that HILIC uses mobile phases that are rich in organic solvent and are thus quite volatile. This helps interface LC systems with mass spectrometers because MS requires that solutes be desolvated and ionized in the gas phase in order to be analyzed. MS detection is desirable because it adds a third dimension to the separation (molecular weight of compound fragments) to the analysis. This additional information is useful

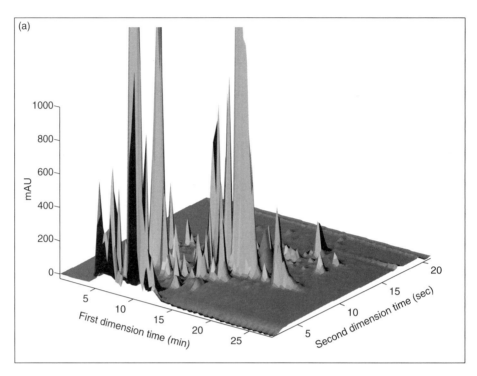

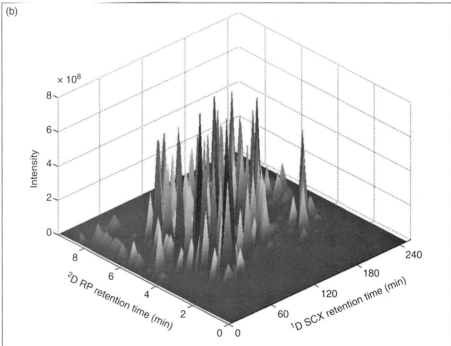

**FIGURE 3.51** Examples of 2D-LC separation. (a) Separation of low-molecular-weight components from a sample of human urine. Discovery HS-F5 (pentafluorophenylpropyl) and ZirChrom-CARB (carbon coated zirconia) phases were used as the two dimensions. (*Source:* From D.R. Stoll, *Anal. Bioanal. Chem.*, 2010, 397, 979–986. Reproduced with permission of Springer.) (b) Separation of peptides from a tryptic digest of bovine serum albumin (BSA). The two dimensions were a strong cation exchange (SCX) and a RPLC column. (*Source:* From R.J. Vonk et al., *Anal. Chem.*, 2015, 87, 5387–5394. Copyright 2015. Reproduced with permission of American Chemical Society.)

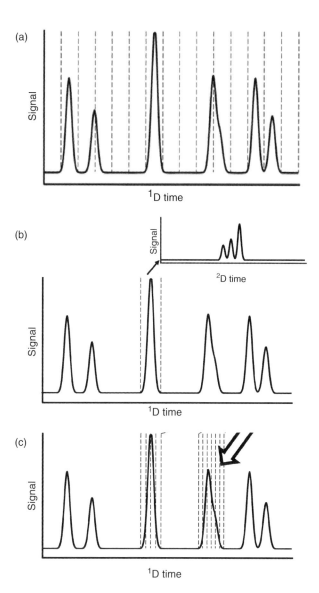

**FIGURE 3.52** Three different sampling modes for 2D-LC. (a) The comprehensive mode (LC×LC) in which the first dimension is sampled at regular time intervals and analyzed on the second dimension column. (b) The heartcutting mode, in which only a single or a very small number of time segments or peaks from the first dimension are targeted for analysis on the second dimension. (c) The selective comprehensive mode, in which multiple regions of the first chromatogram are sampled regularly and analyzed on the second dimension. (*Source:* Reprinted with modification from Groskreutz, S.R. and Stoll, D.R. in *Hydrophilic Interaction Chromatography*, B.A Pack and B.W. Olsen, Eds., 2013, John Wiley and Sons, Inc. Reproduced with permission.)

when screening for specific compound in mixtures that contain hundreds or thousands of chemicals (i.e., finding the proverbial needle in the haystack).

While 2D-LC is not yet routinely practiced in laboratories, its promise of exceptionally high peak capacities, coupled with improvements in commercial instrumentation and the highly desirable ability to use MS detection, will increase its popularity for the comprehensive analysis of complex samples.

## 3.7.   APPLICATION OF LC – ANALYSIS OF PHARMACEUTICAL COMPOUNDS IN GROUNDWATER

A clean, drinkable, and reliable supply of freshwater is critical to supporting the world's population. A number of compounds, including pesticides, household detergents, and biologically active pharmaceuticals such as antibiotics, steroids, stimulants, and hormones, can enter the environment and eventually contaminate the groundwater supply.[78–80] Once in the groundwater, they can make their way into municipal drinking water supplies as indicated in an Associated Press investigation.[81,82] In this section, we focus on the LC analysis of pharmaceutical compounds found in streams across the United States.

One way in which pharmaceutical compounds can enter the freshwater supply is through human excretion coupled with the subsequent release of treated municipal wastewater into rivers and streams. This can occur because treatment plants are not designed to remove water-soluble pharmaceutical compounds. They can also enter the environment through veterinary use in animal feeding when animal waste tanks leak or overflow.

Because these compounds are biologically active, their presence and concentration in the freshwater supply is a concern. The possible consequences of low levels of hormones, steroids, antibiotics, and other pharmaceutical compounds on aquatic systems, animals, and humans are generally unknown but could include the creation of antibiotic-resistant bacteria, increased cancer rates, abnormal physiological processes, and reproductive impairment. One study has linked the presence of birth control drugs such as 17α-ethynylestradiol and 17β-estradiol in rivers to the feminization of male fish.[83,84] In a subsequent study, it was found that in a lake intentionally spiked with 17α-ethynylestradiol, exposure to the drug caused delayed sperm cell development in male fathead minnows in 1 year, and after 2 years the males began producing eggs.[81,85] This caused the population of the minnows to plummet with a concomitant decrease in the lake trout population due to the loss of their food supply. This example shows that it is important to detect and quantify pharmaceutical compounds in rivers and lakes and their possible points of origin.

To address this concern, the U.S. Geological Survey sampled 139 streams across 30 states between 1999 and 2000.[78–80] They focused on streams near large urban centers and livestock production areas (see Figure 3.53 for a map of locations). The study aimed to detect and quantify 95 different organic wastewater contaminants (OWCs) using five different analytical methods. Three of the methods used LC–MS and two were based on GC–MS.

The following sections discuss the methods for sampling, sample treatment, quality assurance, and chromatographic analyses used in these studies. This discussion is intended to show that the sampling and sample treatment aspects of any study are just as important as the actual instrumental analysis of the samples. *Without proper care and concern for the sampling process, the best instrumental methods will not provide meaningful results. Furthermore, without proper attention to the use of controls, blanks, and standards, the data cannot be trusted.*

**FIGURE 3.53** Location of 139 sampling sites in U.S. Geological Survey study of organic contaminants in streams. (*Source:* Reprinted with permission from *Environ. Sci. Technol.*, 2002, 36, 1202–1211. Copyright 2002 American Chemical Society.)

### 3.7.1. Sampling

Streams were selected by their susceptibility to contamination from human, industrial, and agricultural wastewater. They spanned a range in geography, hydrogeology, land use, climate, and basin size. In all, 139 sites were selected.[78–80] Water was sampled at multiple depths at each site and stored in 1-L amber bottles in a refrigerator. Samples were collected in duplicate and people coming in contact with the samples were asked to minimize their use of colognes, perfumes, insect repellant, caffeinated products, and tobacco so as to avoid possible sample contamination.

### 3.7.2. Analysis Method for 21 Antibiotics – Sample Pretreatment

While five different chromatographic methods were used to identify a variety of pharmaceuticals, insecticides, hormones, steroids, and household products, we will focus on a single HPLC method used to identify and quantify 21 different antibiotic compounds.[78] Specifically, this method targeted sulfonamides (SAs), also known as sulfa drugs, and tetracyclines (TCs) (see Figure 3.54). In this method, 500-mL samples of river water were passed through a tandem solid-phase extraction (SPE) cartridge system that included an Oasis hydrophilic–lipophilic-balance (HLB) cartridge and a mixed mode HLB-cation exchange (MCX) cartridge. These cartridges are hollow disks filled with silica particles coated with organic phases. In this way, they are quite similar to the particles commonly used in liquid chromatography. As the sample water is passed through the cartridges, the antibiotics are extracted and concentrated into the organic phase that coats the particles. After the entire 500 mL sample has been passed through the cartridge, the analytes are desorbed

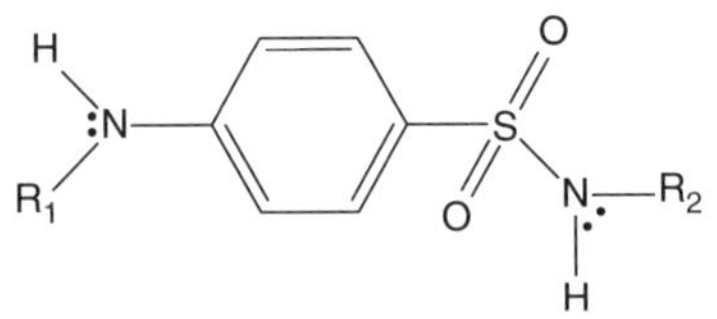

Sulfonamide general structure

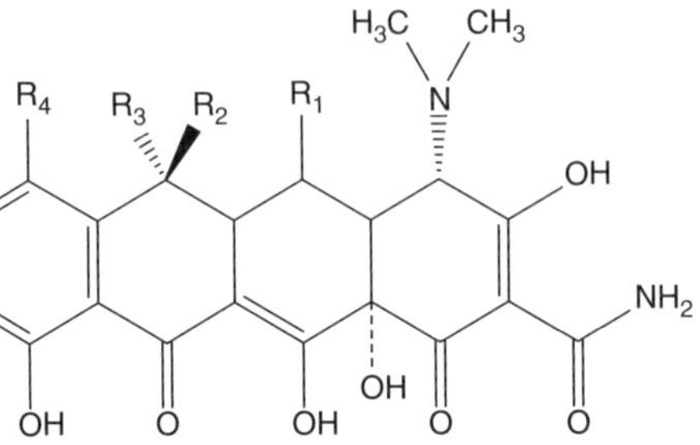

Tetracycline general structure

**FIGURE 3.54** Structure of sulfonamides (SAs), also known as sulfa drugs, and tetracyclines (TCs).[86]

from the extraction phase using a strong organic solvent. In this study, methanol was used to remove the solutes from the HLB cartridge, and a mixture of methanol with 5% $NH_4OH$ was used to elute the antibiotics from the MCX cartridge. The antibiotics are much more soluble in methanol compared to the aqueous solutions from which they were originally extracted. The reason for adding ammonium hydroxide when eluting compounds from the MCX phase is discussed below.

The HLB cartridge extracts organic solutes through dispersion, hydrogen-bonding, dipole–dipole, and dipole–induced dipole interactions. Because it spans this wide range of interactions, it can extract both nonpolar and polar compounds, hence the "hydrophilic-lipophilic" designation. The MCX cartridge adds the ability to retain solutes through specific Coulombic interactions between cationic solutes and the anionic sites on the particles. Because this phase can retain solutes through nonspecific hydrophilic/lipophilic interactions as well as through electrostatic interactions, it is referred to as a mixed-mode phase.

Prior to being passed through these cartridges, the pH of the river samples was adjusted to 3.0 and 0.5 mg of disodium ethylenediaminetetraacetic acid ($Na_2EDTA$) was added. The reason for both of these steps is clear when one considers the general structures of the tetracyclines (TCs) and sulfonamides (SAs) analyzed by this method. The structure of tetracycline indicates that the oxygen-containing moieties can chelate metals through the formation of six-membered rings. The silica to which the extraction phases are bonded contains some metal impurities that cannot be removed. If this is not dealt with, the tetracyclines could irreversibly bond to the metal sites, causing artificially low recoveries of the tetracyclines. For this reason, EDTA, which readily binds to metals, was added to solution. The EDTA occupies the metal sites making them unavailable to the tetracyclines,

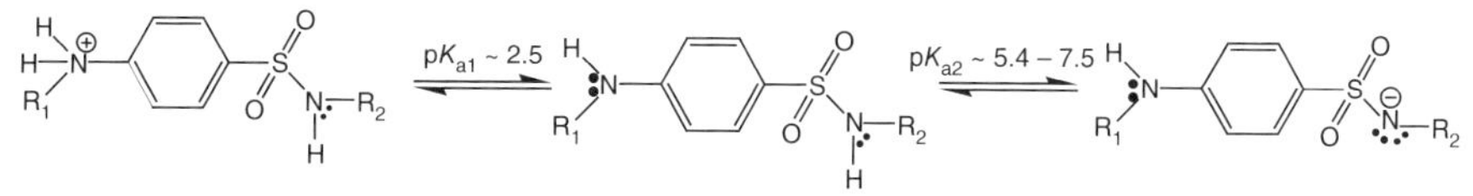

**FIGURE 3.55** Approximate $pK_a$ values for sulfonamides.[86,89]

allowing the tetracyclines to be reversibly absorbed into the organic phase rather than irreversibly bound to the silica particles.

The pH of the water samples was adjusted to 3.0 to force the sulfonamides into a neutral form. There are two acid/base sites of concerns in the general structure of sulfonamides. The $pK_a$ values of these sites vary depending on the specific structure, but they generally fall in the ranges specified in Figure 3.55.[86] At a pH of 3.0, the sulfonamides will generally be in their neutral form and therefore less soluble in water than when they are protonated. This increases their affinity for the hydrophilic/lipophilic extraction phase and thus increases their overall extraction efficiency. At a pH of 3.0, however, some small fraction of the analytes may still be protonated and thus have a net positive charge. Solutes with a positive charge would favor being in the aqueous phase rather than the organic exchange phase and thus would not be extracted from the samples. The cation exchange cartridge (MCX) was present to interact with these positively charged solutes. After the analytes have been extracted from the sample, methanol with $NH_4OH$ is used to desorb the analytes from the MCX phase. The ammonium ions, when introduced into the column, compete for the cation exchange sites and prevent analytes that have desorbed from the phase from readsorbing. This increases the extraction efficiency of the overall process by ensuring that not only neutral solutes but also positively charged compounds are extracted from the stream samples.

### 3.7.3. Use of Internal Standards and Other Quality Assurance Issues

Once the tetracyclines and sulfonamides were removed from the cartridge, the resulting eluate was spiked with 500 ng of $^{13}C_6$-sulfamethezine, which was used as an internal standard for quantification purposes. The volume of eluate was reduced to 20 μL by evaporation of methanol using nitrogen gas and a water bath at 55°C. Then 300 μL of 20 mM aqueous ammonium acetate was added and the resulting solution analyzed by LC–MS. The pharmaceutical compounds that were present in the original 500 mL sample were now present in a volume of only 320 μL, a roughly 1000-fold concentration increase that made it possible to analyze them using the LC–MS methods described as follows.

In addition to these sampling and preconcentration steps, several other steps were taken for quality assurance purposes.[78] For example, ultrapure water blanks, pure water spiked with tetracyclines and sulfonamides, and duplicate samples were extracted via the same procedure to check for potential biases arising from the method. Possible matrix effects were also explored by spiking distilled water, groundwater, and surface water with TCs and SAs and comparing extraction efficiencies in these samples. In addition, several different extraction cartridges were studied to determine which would lead to the best recoveries. These controls and blanks are absolutely essential for obtaining results that can be trusted. They represent careful experimentation and thus provide a good model to follow for experimental design. Consultation of the original studies and related reports for quantifying pharmaceuticals in groundwater[78–80,86–89] is highly encouraged to gain a full appreciation for the careful use of controls, standards, and blanks that were incorporated into these studies.

### 3.7.4.  LC Analyses

After the extraction process, the samples were analyzed by LC–MS using a standard $100 \times 4.6$-mm C-8 column with $3 \mu m$ particles. A binary gradient was used to elute the compounds with a flow rate of $0.6 \, mL/min$. One mobile phase mixture (A) was $10 \, mM$ ammonium formate and 0.3% formic acid in a highly polar 90%/10% water/methanol mixture. The second mobile phase, mixture (B), was $10 \, mM$ ammonium formate with 0.5% formic acid in methanol (i.e., a considerably stronger mobile phase than the almost completely aqueous mobile phase A).

Gradient elution was used starting with 9% B/91% A for 5 min, increasing to 42% B/58% A by 15 min, ending with 100% B by 20 min with a hold at 100% B for 5 min, followed by a 5-min reequilibration time prior to starting the next analysis. Thus, the separation follows a standard reversed-phase mode using a nonpolar stationary phase with a relatively polar mobile phase, combined with gradient elution toward a nonpolar (i.e., stronger) mobile phase to elute strongly retained compounds in a reasonable amount of time.

### 3.7.5.  Mass Spectrometric Selected Ion Monitoring Detection

An electrospray interface operated in the positive ion mode was used to impart charge to the solutes as they eluted from the column. This allowed for detection using a single-quadrupole mass spectrometer. In positive ion mode, the analytes are charged by the addition of a proton to the analyte's original structure, resulting in an $[M+H]^+$ molecular ion peak in the mass spectrum, with possible fragmentation occurring based on the specific structures of the individual analytes.

The distinct molecular weights of each compound, in combination with distinct fragment ions, were used to *identify* the tetracyclines and sulfonamides present in the samples. Furthermore, the ratio of the peak area of the base peak ion of the analytes to the base peak ion of the internal standard was used to *quantify* the amount of each pharmaceutical present (assuming peak heights exceeded the limit of quantitation, which was not always the case).

At times it is desirable to collect the entire mass spectrum of each compound as it elutes from the column. When this is done, the total current resulting from all of the fragments is plotted versus time to yield a chromatogram. In this way, every compound present in the original mixture, regardless of the ions or fragments it produces in the mass spectrometer, is detected and represented by a peak in the chromatogram. When searching for a select class of compounds in potentially complex samples such as groundwater, it can be

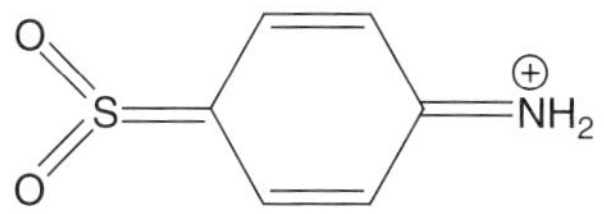

**FIGURE 3.56**   Sulfanilyl fragment ion with $m/z = 156$.[86]

desirable to set the mass spectrometer to a fixed mass-to-charge ratio and detect only those compounds that produce ions with the specific ratios of interest. For example, it is common for sulfonamides, regardless of their individual structural differences, to fragment into the ion shown in Figure 3.56 that has a mass-to-charge ratio of 156.[86] In this way, the analysis can be made selective for sulfonamides by selecting $m/z = 156$ as the only ion detected by the mass spectrometer. Thus, rather than continuously acting in a fast scan mode and collecting the entire mass spectrum at every point in time, the detector is operated in selected ion monitoring (SIM) mode. In this case, only solutes that produce a fragment with an $m/z$ ratio of 156 are detected and represented by a peak in the chromatogram. This considerably simplifies the chromatogram and reduces the potential overlap of signals that could complicate quantitative efforts. Of the five methods used to detect a total of 95 OWCs, all five used mass spectrometers as the detector, and two were able to take advantage of the specificity of SIM.

### 3.7.6.   Results

Using the method described in the preceding section in combination with four other methods, organic wastewater contaminants were found in 80% of the streams sampled for this study. Of the 95 OWCs that were tested for, 82 were detected at least once. Some of the most frequently detected compounds included coprostanol (steroid), cholesterol (steroid), N,N-diethyltoluamide (DEET insect repellant), caffeine (stimulant), triclosan (antimicrobial disinfectant), and 4-nonylphenol (nonionic detergent metabolite).[78]

The concentrations of OWCs were generally low ($<1\,\mu g/L$) and seldom rose above guidelines for acceptable levels for aquatic life and for drinking water. Furthermore, it is difficult, if not impossible, to draw conclusions from this study alone about the possible environmental and health effects of continuous exposures to low levels of OWCs. In that regard, while the effects of these OWCs are not completely known, this study provides valuable information regarding the current levels of these contaminants. These studies can be compared to future studies to provide long-term trends that could then be correlated with possible health and environmental trends.

Just as the men and women at the U.S. Customs and Border Protection Laboratory deserve credit for their dedication to providing accurate and useful analysis of oil and a broad range of other imported materials (see Chapter 2), the men and women of the U.S. Geological Survey deserve credit for undertaking this sweeping investigation of our critically important supply of freshwater. This study illustrates that our health and that of the aquatic environment depends on their efforts and on having trained scientists skilled in the use of instrumental methods of analysis, including HPLC and mass spectrometry.

## 3.8. SUMMARY

Liquid chromatographic separations are based on the relative strength of the intermolecular interactions between solutes and the stationary and mobile phases. The most common mode is RPLC, which uses a nonpolar stationary phase in combination with a polar mobile phase. RPLC columns are typically packed with small (micrometer-sized) porous silica particles to which alkyl chains are covalently linked. In addition to RPLC, many other modes of chromatography exist, such as NPLC, SEC, IEX, HILIC, and affinity chromatography. These modes take advantage of specific properties of the solutes such as their size or charge to achieve separations. Regarding instrumentation, a typical LC system includes solvent reservoirs, a pump, an injector, the column, and a detector. Many methods are used to detect molecules as they elute from the column. The most common is UV–visible absorbance detection, but other detection methods such as mass spectrometry in particular, as well as electrochemistry and fluorimetry, are also used routinely. In this chapter, we discussed applications of LC to drug testing, food safety, chiral analyses, and the analysis of pharmaceutical compounds in environmental samples, but this is just a very small subset of all of the areas of application. Because of the ability of liquid chromatography to separate and quantify components in mixtures, it is arguably one of the most widely used instrumental techniques in all areas of science.

## PROBLEMS

3.1 Predict the order of elution of the following compounds on a C-18 stationary phase with a typical reversed-phase mobile phase (i.e., aqueous with some organic modifier): 4-ethylphenol, *N*,*N*-diethylaniline, aniline, anthracene. Draw the structures of each and explain your reasoning by considering the interactions of each solute with both an aqueous mobile phase and a C-18 stationary phase.

3.2 Which mode of liquid chromatography would be best for separating the following?
   (a) A mixture of high molecular weight polystyrenes
   (b) Fluoride and nitrate in water
   (c) Anti-rabbit IgG in a mixture of other proteins
   (d) The active ingredients in cough medicine
   (e) Large planar polyaromatic hydrocarbons
   (f) A mixture of *R*- and *S*-enantiomers of a new drug to decrease cholesterol.

3.3 Which ions could be used in the mobile phase to elute $Zn^{2+}$ in ion-exchange chromatography?

3.4 Consider the following mixture of acids and bases and their respective $pK_a$'s or $pK_b$'s.
   (a) Predict the order of elution in IEX using a sulfonate stationary phase in mobile phases of pH 3.8 and 8. Some solutes may coelute at various pHs.

(b) Why isn't there a $pK_b$ value listed for sodium butylsulfate?

(c) Why is aniline such a weak base compared to diethylamine?

| | |
|---|---|
| Aniline | $pK_b = 9.4$ |
| Diethylamine | $pK_b = 3.3$ |
| Benzoic acid | $pK_a = 4.2$ |
| Sodium butylsulfate | |
| Chloroethanoic acid | $pK_a = 2.9$ |

**3.5** Does the mobile phase flow through typical RPLC spherical porous particles? If not, how do solutes get into the pores to be retained? How do they get out of the pores? Contrast this with mobile phase flow in monolithic columns.

**3.6** Suppose that urea, a very small molecule, has a retention volume of 28.0 mL in a size exclusion column. Do you expect $V_R$ for the following solutes to be much larger than, larger than, approximately equal to, smaller than, or much smaller than the $V_R$ of urea?

(a) The tobacco mosaic virus

(b) A polyaromatic hydrocarbon with the formula $C_{78}H_{26}$

(c) Methanol

(d) Polystyrene polymers around 45,000 Da.

**3.7** Typical RPLC silica particles should not be used with mobile phases with a pH below 2 or above 8. What happens at each of these extremes that degrades the particles or stationary phase?

**3.8** Suppose you wish to separate 4-nitrophenol from 3-nitrophenol. A separation using a 30:70 water/THF mobile phase on a 10 cm × 4.6 mm C-8 column with 5 μm particles is not providing the desired separation. List four *specific* changes you could make to improve the resolution of the separation. Explain your reasoning for each and why you expect them to improve the separation. Of the ones you suggested, which is the easiest and cheapest and therefore the thing to be tried first?

**3.9** Figure 3.34 compares the response of an ELSD detector to that of a UV–visible detector for isoflavones and saponins. Find the structures of genistein (an isoflavone) and soyasaponin I (a saponin). What structural features does soyasaponin I lack that normally make compounds absorb UV–visible light?

**3.10** What detector could be used to detect the following?

(a) $Ca^{2+}$

(b) Octane

(c) Anthracene

(d) A high-molecular-weight polymer

(e) A new pharmaceutical compound about which the molecular weight and structural information is desired

(f) A mixture of neurotransmitters.

**3.11** Why is ion suppression so valuable in detecting ions when using IEX?

**3.12**   (a) Draw the structures of L- and D-dopa. (Hint: the Internet will likely be helpful in answering parts of this question).

   (b) Which is the chiral carbon?

   (c) Which form is biologically active and what is it used to treat medically?

   (d) Suggest a type of column that could be used to separate L- and D-dopa.

**3.13**   A typical HPLC column is 15.0 cm long and 4.6 mm in diameter.

   (a) What is the internal volume of the column?

   (b) What is the volume of a 5 µm diameter particle (approximate it as a hard sphere)?

   (c) Because of the way spherical particles pack in a column, the volume that is external to the particles is approximately 40% of the entire column volume, meaning that 60% of the column volume is occupied by the particles. Given this, the total column volume you calculated in part (a), and the volume of a particle you calculated in b, calculate the number of particles packed in a 15.0 cm × 4.6 mm column.

**3.14**   What is endcapping and why is it done?

**3.15**   In this chapter, we equated an average solute with the size of a pea and found that a single 5 µm particle would be approximately 560 ft in diameter. To extend the analogy and keeping on this scale, what would be the length and diameter of a typical 15.0 cm × 4.6 mm in miles? How does this compare to the distance between New York City and Los Angeles (2500 miles)? Think about how a solute the size of a pea compares to the total size of this column.

**3.16**   In the USGS study, why were the stream samples stored in amber bottles and refrigerated? If they had been stored in clear bottles, would the quantitative results for any individual OWC likely be artificially high or artificially low?

**3.17**   Why is a C-13 labeled sulfonamide a good choice for an internal standard for the LC–MS analysis of pharmaceuticals in groundwater?

**3.18**   Many separations can be achieved in 30 min on a standard RPLC column (15.0 cm × 4.6 mm) operated at a flow rate of 1.5 mL/min. Many fast HPLC separations on narrow bore columns can be achieved in approximately 5 min at flow rates of 400 µL/min.

   (a) If both systems are operated continuously for 1 year, what is the cost savings of using the narrow bore column assuming it costs $30.00 to dispose of 1 L of waste mobile phase?

   (b) How many analyses are performed by both systems in 1 year?

   (c) What can you conclude from your answers to parts (a) and (b)?

## REFERENCES

1. Lock, S., in *High-Throughput Analysis for Food Safety*, Wang, P.G.; Vitha, M.F.; and Kay, J.F. (Eds.); John Wiley & Sons, Inc.: Hoboken, 2014.

2. Sudan Dyes I to IV in Food, BfR opinion, Nov. 2003. http://www.bfr.bund.de/cm/349/dyes_sudan_I_IV.pdf, last accessed July 7, 2015.

3. Koupparis, M.A.; Megoulas, N.C.; Gremilogianni, A.M., in *Hydrophilic Interaction Chromatography: A Guide for Practitioners*. Olsen, B. and Pack, B. (Eds.); John Wiley & Sons, Inc.: Hoboken, 2013.

4. Ihunegbo, F.N.; Tesfalidet, S.; Jiang, W. *J. Sep. Sci.* 2010, *33*, 988–995.

5. Davies, E. Chemistry World, 2012 http://www.rsc.org/chemistryworld/2012/06/chemistry-olympics, last accessed July 7, 2015.

6. Jeong, E.S.; Kim, S.-H.; Cha, E.-J.; Lee, K.M.; Kim, H.J.; Lee, S.-W.; Kwon, O.-S.; Lee, J. *Rapid Commun. Mass. Spectrom.* 2015, *29*, 367–384.

7. Dominguez-Romero, J.C.; Garcia-Reyes, J.F.; Lara-Ortega, F.J.; Molina-Diaz, A. *Talanta*, 2015, *134*, 74–88.

8. World Anti-Doping Agency website, Technical documents, TD2014EPO-summary modifications, Sept. 19, 2014. https://www.wada-ama.org/en/resources/science-medicine/td2014-epo-summary-modifications, accessed July 7, 2015.

9. Ettre, L.S. *LC/GC*, 2003, *21*, 458–467.

10. Giddings, J.C. *Dynamics of Chromatography: Principles and Theory*; Marcel Dekker: New York, 1965.

11. Giddings, J.C. *Unified Separation Science*; John Wiley & Sons: New York, 1991.

12. Ettre, L.S. *LC/GC* 2005, *23*, 486–495.

13. Meyer, V.R. *Practical High-Performance Liquid Chromatography*, 4th Ed.; Wiley-VCH Verlag GmbH & Co.: Weinheim, 2004.

14. Unger, K.K. *Porous Silica, Its Properties and Use as Support in Column Liquid Chromatography*; Elsevier Scientific Publishing Co.: Amsterdam, 1979.

15. Snyder, L.R.; Kirkland, J.J. *Introduction to Modern Liquid Chromatography*, 2nd Ed.; John Wiley & Sons: New York, 1979.

16. Yu, Y.-X.; Gao, G.-H. *Fluid Phase Equilib.* 1999, *166*, 111–124.

17. Schure, M.R.; Rafferty, J.L.; Zhang, L.; Siepmann, J.I., *LC/GC North Am.* 2013, *31*, 630–637.

18. Lubda, D.; Cabrera, K.; Kraas, W.; Schaefer, C.; Cunningham, D. *LC/GC* 2001, *19*, 1186–1191.

19. Majors, R.E. *LC/GC* 2000, *18*, 1214–1227.

20. Svec, F.; Huber, C.G. *Anal. Chem.* 2006, *78*, A2100–A2107.

21. Jacoby, M. *Chem. Eng. News* 2006, *84* (50), 14–19.

22. Liteanu, C.; Gocan, S.; *Gradient Liquid Chromatography*, Chalmers, R.A. (Eds.); John Wiley & Sons, Inc.: New York, 1974.

23. Ettre, L.S. *LC/GC* 2004, *22*, 514.

24. Snyder, L.R. *J. Chromatogr. Sci.* 1978, *16*, 223–234.

25. Snyder, L.R.; Carr, P.W.; Rutan, S.C. *J. Chromatogr. A* 1993, *656*, 537–547.

26. Majors, R.E.; Przybyciel, M. *LC/GC* 2002, *20*, 584–593.

27. Wirth, M. J.; Fatunmbi, H.O. *Anal. Chem.* 1992, *64*, 2783–2786.

28. Wirth, M. J.; Fatunmbi, H.O. *Anal. Chem.* 1993, *65*, 822–826.

29. Sander, L.C.; Pursch, M.; Wise, S.A. *Anal. Chem.* 1999, *71*, 4821–4830.

30. Lucy, C. *J. Chromatogr. A* 2003, *1000*, 711–724.

31. Fritz, J. *J. Chromatogr. A* 2004, *1039*, 3–12.

32. Small, H. *J. Chem. Educ.* 2004, *81*, 1277–1284.

33. Small, H.; Stevens, T.S.; Bauman, W.C. *Anal. Chem.* 1975, *47*, 1801–1809.

34. Small, H. *Ion Chromatography*; Plenum Press: New York, 1989.

35. Walton, H.F.; Rocklin, R.D. *Ion Exchange in Analytical Chemistry*; CRC Press: Boca Raton, 1990.

36. Alpert, A.J. *J. Chromatogr.* 1990, *499*, 177–196.

37. Boguslaw, B.; Moga, S. *Anal. Bioanal. Chem.* 2012, *402*, 231–247.

38. Neue, U.D. http://www.sepscience.com/Techniques/LC/Articles/501-/Hydrophilic-interaction-chromatography-HILIC, accessed July 16, 2015.

39. Borman, S.A. *Anal. Chem.* 1983, *55*, 384A–390A.

40. Tosoh Bioscience, Size Exclusion Chromatography Catalog, www.separations.us.tosohbioscience.com, accessed March 7, 2016.

41. Walters, R.R. *Anal. Chem.* 1985, *57*, 1099A–1114A.

42. Mohr, P. *Affinity Chromatography; Practical and Theoretical Aspects*; CRC Press: Boca Raton, 1985.

43. Poole, C.F.; Poole, S.K. *Chromatography Today*; Elsevier Science Publishers: Amsterdam, 1991.

44. http://www.chromacademy.com/resolver-september2010_High_Efficiency_HPLC_Separations.html, last accessed July 14, 2015.

45. Thompson, J.D.; Brown, J.S.; Carr, P.W. *Anal. Chem.* 2001, *73*, 3340–3347.

46. Majors, R.E. *LC/GC*, 2005, *23*, 1248–1255.

47. Cole, R.B. *Electrospray Ionization Mass Spectrometry*; John Wiley & Sons, Inc.: New York, 1997.

48. Yamashita, M.; Fenn, J.B. *J. Phys. Chem.* 1984, *88*, 4451–4459.

49. Hu, Q.; Noll, R.J.; Li, H.; Makarov, A.; Hardman, M.; Cooks, R.G. *J. Mass Spectrom.* 2005, *40*, 430–443.

50. Kaufmann, A. *Anal. Bioanal. Chem.* 2012, *403*, 1233–1249.

51. Lindsay, S. *High Performance Liquid Chromatography*; John Wiley & Sons, Inc.: Chichester, 1987.

52. Eriksson, T.; Bjorkman, S.; Roth, B.; Fyge, A.; Hoglund, P. *Chirality* 1995, *7*, 44–52.

53. http://www.k-faktor.com/thalidomide/, accessed March 7, 2016.

54. von Moos, R.; Stolz, R.; Cerny, T.; Gillessen, S. *Swiss Med. Wkly.* 2003, *133*, 77–87.

55. Welch, C.; Szczerba, T.; Perrin, S. *Review of Chiral Separations*, Regis Technology, http://www.registech.com/chiral/chiralreview.pdf, last accessed March 07, 2016.

56. Welch, C.J. *J. Chromatogr. A* 1994, *666*, 3–26.

57. Pirkle, W.H.; House, D.W.; Finn, J.M. *J. Chromatogr.*, 1980, *192*, 143–158.

58. Pirkle, W.H.; Finn, J.M. *J. Org. Chem.* 1981, *46*, 2935–2938.

59. Pirkle, W.H.; Tsipouras, A. *J. Chromatogr.* 1984, *291*, 291–298.

60. Pirkle, W.H.; Burke, III, J.A. *J. Chromatogr.* 1991, *557*, 173–185.

61. Fujimura, K.; Ueda, T.; Ando, T. *Anal. Chem.* 1983, *55*, 446–450.

62. Kawaguchi, Y., Tanaka, M., Nakae, M., Funazo, K., Shono, T. *Anal. Chem.* 1983, *55*, 1852–1857.

63. Armstrong, D.W., DeMond, W. *J. Chromatogr. Sci.* 1984, *22*, 411–415.

64. Armstrong, D.W.; Tang, Y.; Chen, S.; Zhou, Y.; Bagwill, C.; Chen, J.R. *Anal. Chem.* 1996, *66*, 1473–1484.

65. Chen, S.; Liu, Y.; Armstrong, D.W.; Borrell, J.I.; Martinez-Teipel, B.; Matallana, J.L. *J. Liq. Chromatogr.* 1995, *18*, 1495–1507.

66. Berthod, A.; Liu, Y.; Bagwill, C.; Armstrong, D.W. *J. Chromatogr. A* 1996, *731*, 123–137.

67. Majors, R.E. *LC/GC* 2004, *22*, 416–428.

68. Wehr, T. *LC/GC* 2002, *20*, 40–47.

69. Carr, P.W.; Wang, X.; Stoll, D.R. *Anal Chem.* 2009, *81*, 5342–5353.

70. Carr, P.W.; Wang, X.; Stoll, D.R. *LC/GC North Am.* 2010, *28*, 932–942.

71. Mao, Y.; Carr, P.W. *LC/GC* 2003, *21*, 150–167.

72. Mao, Y.; Carr, P.W. *Anal. Chem.* 2000, *72*, 110–118.

73. Bushey, M.M.; Jorgenson, J.W. *Anal. Chem.* 1990, *62*, 161–167.

74. Liang, Z.; Li, K.; Wang, X.; Ke, Y.; Jin, Y.; Liang, X. *J. Chromatogr. A* 2012, *1224*, 61–69.

75. Stoll, D.R. *Anal. Bioanal. Chem.* 2010, *397*, 979–986.

76. Vonk, R.J.; Gargano, A.F.G.; Davydova, E.; Dekker, H.L.; Eeltink, S.; de Koning, L.J.; Schoenmakers, P.J. *Anal. Chem.* 2015, *87*, 5387–5394.

77. Groskreutz, S.R.; Stoll, D.R. in *Hydrophilic Interaction Chromatography: A Guide for Practitioners*, Olsen, B.A. and Pack, B.W. (Eds.); John Wiley & Sons, Inc.: Hoboken, 2013.

78. Kolpin, D.W.; Furlong, E.T.; Meyer, M.T.; Thurman, E.M.; Zaugg, S.D.; Barber, L.B.; Buxton, H.T. *Environ. Sci. Technol.* 2002, *36*, 1202–1211.

79. Barnes, K.K.; Kolpin, D.W.; Meyer, M.T.; Thurman, E.M.; Furlong, E.T.; Zaugg, S.D.; Barber, L.B. U.S. Geological Survey Open-File Report 02-92. http://toxics.usgs.gov/pubs/OFR-02-94/, accessed March 07, 2016.

80. Buxton, H.T.; Kolpin, D.W. U.S. Geological Survey Fact Sheet FS-027-02. http://toxics.usgs.gov/pubs/FS-027-02/, accessed March 07, 2016.

81. Halford, B. *Chem. Eng. News* 2008, *86* (8), 13–17.

82. Donn, J.; Mendoza, M.; Pritchard, J. *PharmaWater I–III, Associated Press archives three part series*, March 9–11, 2008.

83. Desbrow, C.; Routledge, E.J.; Brighty, G.C.; Sumpter, J.P.; Waldock, M. *Environ. Sci. Technol.* 1998, *32*, 1549–1558.

84. Routledge, E.J.; Sheahan, D.; Desbrow, C.; Brighty, G.C.; Waldock, M.; Sumpter, J.P. *Environ. Sci. Technol.* 1998, *32*, 1559–1565.

85. Kidd, K.A.; Blanchfield, P.J.; Mills, K.H.; Palace, V.P.; Evans, R.E.; Lazorchak, J.M.; Flick, R.W. *Proc. Natl. Acad. Sci. USA* 2007, *104*, 8897–8901.

86. Lindsey, M.E.; Meyer, M.; Thurman, E.M. *Anal. Chem.* 2001, *73*, 4640–4646.

87. Hirsch, R.; Ternes, T.A.; Haberer, K.; Mehlich, A.; Ballwanz, F.; Kratz, K.-L. *J. Chromatogr. A* 1998, *815*, 213–223.

88. Göbel, A.; McArdell, C.S.; Suter, M. J.-F.; Giger, W. *Anal. Chem.* 2004, *76*, 4756–4764.

89. Huang, C.-H.; Report for 2002GA11B: Investigation of Chlorination and Ozonation of Antibiotics in Georgia Water. http://water.usgs.gov/wrri/02grants/prog-compl-reports/2002GA11B.pdf.

# SOLUTIONS

## FUNDAMENTALS OF CHROMATOGRAPHY

**1.1** (a) The structures are as follows:

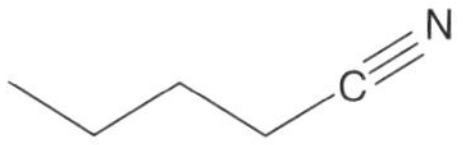

Valeronitrile

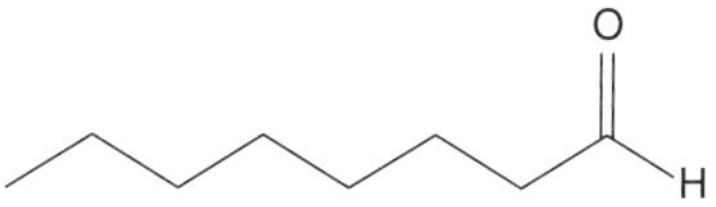

Octanal

Both have polar moieties that will interact with water via dipole–dipole and hydrogen-bonding interactions. The major difference, however, is that octanal has a longer hydrophobic carbon chain, which will make it preferentially interact with hexadecane more so than the valeronitrile. This is the main reason that octanal has a larger distribution coefficient.

(b) First do the calculation for valeronitrile:

$$K = \frac{[\text{valeronitrile}]_{C16}}{[\text{valeronitrile}]_{\text{water}}} = \frac{\left(\frac{n_{C16}}{V_{C16}}\right)}{\left(\frac{n_{\text{water}}}{V_{\text{water}}}\right)} = 3.236$$

The abbreviation "C16" is being used to represent the hexadecane phase.

$$n_{C16} + n_{\text{water}} = n_{\text{total}}$$

$$n_{C16} + n_{\text{water}} = 1,000,000 \quad \text{so} \quad n_{C16} = 1,000,000 - n_{\text{water}}$$

*Chromatography: Principles and Instrumentation*, First Edition. Mark F. Vitha.
© 2017 John Wiley & Sons, Inc. Published 2017 by John Wiley & Sons, Inc.

We know $V_{C16} = V_{water}$, so the volumes cancel. Substituting yields

$$K = \frac{1,000,000 - n_{water}}{n_{water}} = 3.236$$

$$1,000,000 = 3.236\, n_{water} + n_{water} = 4.236\, n_{water}$$

$$2.361 \times 10^5 = n_{water} \text{ for valeronitrile}$$

Following the same procedure yields ~2158 molecules of octanal in the aqueous phase at equilibrium, meaning the other 997,842 octanal molecules are in the hexadecane.

(c) The same procedure is used, except that $n_{total}$ is now 23,610 for valeronitrile and 2158 for octanal because only the water phase was moved, meaning some molecules were left behind in the hexadecane stationary phase. Applying the same method yields

$n_w = 55,740$ and $n_{C16} = 180,360$ for valeronitrile and
$n_w = 4.66$ (or about five molecules) and $n_{C16} = 2153$ for octanal.

(d) What had been an equal mixture of both valeronitrile and octanal is now essentially separated, with much more valeronitrile in the aqueous phase and much more octanal in the stationary phase (consistent with the relative magnitudes of their distribution coefficients).

(e) If repeated, the water phase would eventually contain virtually no octanal molecules, meaning that valeronitrile would be enriched in that phase relative to octanal. The octanal is accumulating in the stationary phase.

(f) Again we'll first do the calculation for valeronitrile:

$$K = \frac{[\text{valeronitrile}]_{C16}}{[\text{valeronitrile}]_{water}} = \frac{\left(\frac{n_{C16}}{V_{C16}}\right)}{\left(\frac{n_{water}}{V_{water}}\right)} = \frac{\left(\frac{n_{C16}}{50.00\,\text{mL}}\right)}{\left(\frac{n_{water}}{950.0\,\text{mL}}\right)} = \frac{n_{C16}}{n_{water}}\left(\frac{950.0\,\text{mL}}{50.00\,\text{mL}}\right) = 3.236$$

$$\frac{n_{C16}}{n_{water}} = 3.236\left(\frac{50.0\,\text{mL}}{950.00\,\text{mL}}\right) = 0.1703$$

$$n_{C16} + n_{water} = n_{total}$$

$$n_{C16} + n_{water} = 1,000,000 \quad \text{so} \quad n_{C16} = 1,000,000 - n_{water}$$

$$\frac{1,000,000 - n_{water}}{n_{water}} = 0.1703$$

$$1,000,000 = 0.1703\, n_{water} + n_{water} = 1.1703\, n_{water}$$

$$8.545 \times 10^5 = n_{water} \text{ for valeronitrile}$$

There is a lot more left in the water compared to when the volumes were equal ($n_{water}$ was $2.361 \times 10^5$).

Repeating for octanal yields $n_{water} = 39,470$ (compared to 2158 when the volumes were equal).

More molecules of both solutes are found in the mobile phase because of the increase in the phase ratio, illustrating the effect that the phase ratio has on partitioning.

**1.2**  (a)

$$\text{Dichlorvos:} \quad k = \frac{t_r - t_m}{t_m} = \frac{22.84\,s - 3.80\,s}{3.80\,s} = 5.01$$

$$\text{Mevinphos:} \quad k = \frac{t_r - t_m}{t_m} = \frac{31.53\,s - 3.80\,s}{3.80\,s} = 7.30$$

(b) For dichlorvos, there are five molecules in the stationary phase for each one in the mobile phase. For mevinphos, there are 7.30 molecules in the stationary phase for each one in the mobile phase (or approximately 22 in the stationary phase for every 3 in the mobile phase).

(c) The mevinphos spends more time in the stationary phase and therefore has a longer retention time and higher retention factor.

(d) Both spend the same amount of time in the mobile phase. Recall that time is velocity/distance and all species travel the same distance – the length of the column – and all move down the column at the same velocity – the velocity of the mobile phase. Because the distance and the velocity are the same for all solutes, all solutes spend the same amount of time in the mobile phase. Separation therefore occurs because of differences in the time spent in the stationary phase.

**1.3**

0.530 mm (Radius of column)
2

$5.00 \times 10^{-6}$ m film thickness

30.0 m length

$$V_{total} = \pi r^2 L = \pi \left( \frac{0.530 \times 10^{-3}\,m}{2} \right)^2 (30.0\,m) = 6.619 \times 10^{-6}\,m^3$$

$$V_{mobile} = \pi \left( \frac{0.530 \times 10^{-3}\,m}{2} - 5.00 \times 10^{-6}\,m \right)^2 (30.0\,m) = 6.371 \times 10^{-6}\,m^3$$

$$V_{stationary} = V_{total} - V_{mobile} = 6.619 \times 10^{-6}\,m^3 - 6.371 \times 10^{-6}\,m^3 = 2.48 \times 10^{-7}\,m^3$$

$$\beta = \frac{V_{mobile}}{V_{stationary}} = \frac{6.371 \times 10^{-6}\,m^3}{2.48 \times 10^{-7}\,m^3} = 25.7$$

**1.4**  Using a column with a thicker film leads to longer retention times for all solutes. Mathematically, this can be understood because $k = K/\beta$ and $\beta = V_{mobile}/V_{stationary}$. So $k = K * V_{stationary}/V_{mobile}$. As $V_{stationary}$ increases, $k$ increases. But $k = (t_r - t_m)/t_m$. Increasing the stationary phase thickness does not affect the dead time ($t_m$), which is set by the velocity and the column length. So the only way that $k$ can increase is if $t_r$ increases. So as the stationary phase thickness increases, retention increases for all solutes.

**1.5**
(a) $\alpha = \dfrac{k_{iodobenzne}}{k_{bromobenze}} = \dfrac{\dfrac{4.57\,min - 1.56\,min}{1.56\,min}}{\dfrac{3.87\,min - 1.56\,min}{1.56\,min}} = 1.30$

(b) The separation factor would decrease as the retention of iodobenzene decreased toward that of bromobenzene. The two peaks would get closer together, indicating less selectivity in the separation.

**1.6**
$$N = 5.54\left(\dfrac{17.456\,min}{0.276\,min}\right)^2 = 22,160\,plates$$

$$H = \dfrac{L}{N}$$

$$H = \dfrac{25.0\,cm}{22,160\,plates} = 0.00113\,cm/plate$$

**1.7**  Slow mass transfer in the mobile phase contributes more significantly than longitudinal diffusion at high linear velocity. Longitudinal diffusion contributes more significantly at low linear velocity.

**1.8**  One mechanism is called the flow mechanism that relaxes broadening because in any single path through the column, there will be regions that are more tightly packed that restrict and slow the flow, and other regions that are more open, with higher velocities. This causes some averaging of the velocity even within a single path, which reduces band broadening overall.

   The second mechanism is the diffusion mechanism. This mechanism describes the fact that solutes diffuse radially from one path into another, and thereby experience the average velocity of a number of paths. The averaging reduces the broadening that would otherwise occur if solutes stayed in just a single flow path through the column.

**1.9**
(a) $H_2$ Faster diffusion of solutes in $H_2$ decreases broadening at flow rates above the optimum – so faster separations can be conducted without significant increases in broadening. While $H_2$ is better than $N_2$, it is not frequently used in practice because of the hazards associated with $H_2$. Instead, helium is commonly used because its behavior is close to that of $H_2$ but without the potential for explosions.

(b) $d_p = 2.00\,\mu m$ – smaller particles lead to less broadening.

(c) $0.25\,\mu m$ film thickness. Thinner films mean the solute has less distance to travel in the stationary phase before it can diffuse back into the mobile phase and catch up to the solutes that stayed in the mobile phase. This reduces broadening (it also reduces the net retention as illustrated in other problems).

(d) Decrease the linear velocity to reduce broadening when one is above the optimum value. This reduces the contribution to broadening incorporated in the C-terms.

**1.10**  For He conditions:

(a)

| $\bar{u}$ (cm/s) | First term (cm) | Second term (cm) | Third term (cm) | $H$ (cm) |
|---|---|---|---|---|
| 2 | 0.19100 | 0.00052 | 0.000012 | 0.1915 |
| 5 | 0.07640 | 0.00129 | 0.000031 | 0.0777 |
| 10 | 0.03820 | 0.00258 | 0.000062 | 0.0408 |
| 15 | 0.02547 | 0.00387 | 0.000094 | 0.0294 |
| 20 | 0.01910 | 0.00516 | 0.000125 | 0.0244 |
| 25 | 0.01528 | 0.00645 | 0.000156 | 0.0219 |
| 30 | 0.01273 | 0.00774 | 0.000187 | 0.0207 |
| 35 | 0.01091 | 0.00903 | 0.000219 | 0.0202 |
| 40 | 0.00955 | 0.01032 | 0.000250 | 0.0201 |
| 50 | 0.00764 | 0.01290 | 0.000312 | 0.0208 |
| 60 | 0.00637 | 0.01548 | 0.000375 | 0.0222 |
| 70 | 0.00546 | 0.01806 | 0.000437 | 0.0240 |
| 80 | 0.00478 | 0.02064 | 0.000500 | 0.0259 |
| 90 | 0.00424 | 0.02322 | 0.000562 | 0.0280 |
| 100 | 0.00382 | 0.02579 | 0.000625 | 0.0302 |

(b)  and (c)

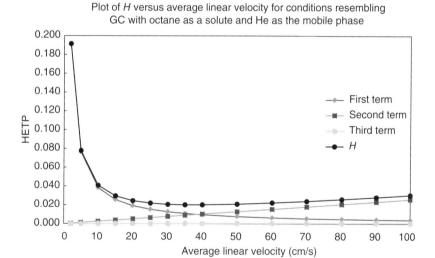

Plot of $H$ versus average linear velocity for conditions resembling GC with octane as a solute and He as the mobile phase

(d)  Optimum velocity (minimum $H$) occurs between 35 and 40 cm/s under the conditions specified.

(e)

| $\bar{u}$ (cm/s) | First term (cm) | Second term (cm) | Third term (cm) | $H$ (cm) |
|---|---|---|---|---|
| 2 | 0.04600 | 0.00214 | 0.000012 | 0.0482 |
| 5 | 0.01840 | 0.00536 | 0.000031 | 0.0238 |
| 10 | 0.00920 | 0.01071 | 0.000062 | 0.0200 |
| 15 | 0.00613 | 0.01607 | 0.000094 | 0.0223 |
| 20 | 0.00460 | 0.02142 | 0.000125 | 0.0261 |
| 25 | 0.00368 | 0.02678 | 0.000156 | 0.0306 |
| 30 | 0.00307 | 0.03213 | 0.000187 | 0.0354 |
| 35 | 0.00263 | 0.03749 | 0.000219 | 0.0403 |
| 40 | 0.00230 | 0.04284 | 0.000250 | 0.0454 |
| 50 | 0.00184 | 0.05355 | 0.000312 | 0.0557 |
| 60 | 0.00153 | 0.06426 | 0.000375 | 0.0662 |
| 70 | 0.00131 | 0.07497 | 0.000437 | 0.0767 |
| 80 | 0.00115 | 0.08568 | 0.000500 | 0.0873 |
| 90 | 0.00102 | 0.09639 | 0.000562 | 0.0980 |
| 100 | 0.00092 | 0.10710 | 0.000625 | 0.1086 |

Plot of $H$ versus average linear velocity for conditions resembling GC with octane as a solute and $N_2$ as the mobile phase

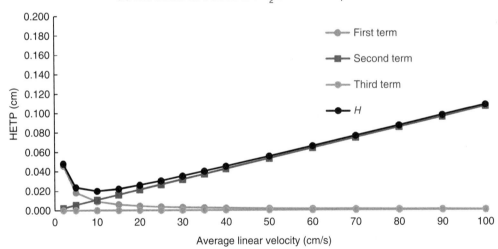

Comparison of $H$ with He versus $N_2$ mobile phase

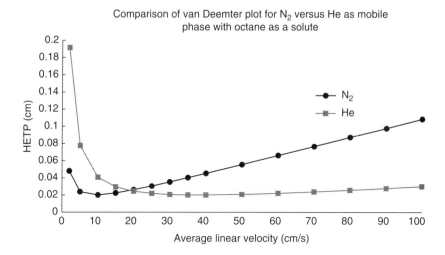

Comparison of van Deemter plot for $N_2$ versus He as mobile phase with octane as a solute

The optimum velocity is around $10\,\text{cm/s}$ in $N_2$, whereas it is near $30\text{--}35\,\text{cm/s}$ in He under the conditions specified. So with He, the solutes can be eluted much faster (shorter analysis time) without significantly more broadening. Furthermore, $H$ does not increase at high velocities much with He as the mobile phase, so higher velocities can be used with He without significantly broadening the peaks and risking peak overlap. With $N_2$, $H$ rises rapidly with increasing flow rates due to the influence on the diffusion coefficient in the C-term.

**1.11**

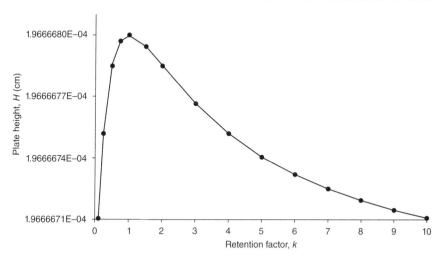

**1.12** An asymmetry factor less than 1.00 indicates that the peak is fronted.

**1.13** (a) $R = \dfrac{t_{r,B} - t_{r,A}}{\frac{1}{2}(W_B + W_A)} = \dfrac{7.028\,\text{min} - 6.912\,\text{min}}{\frac{1}{2}(0.099\,\text{min} + 0.095\,\text{min})} = \dfrac{0.116}{0.097} = 1.196$

(b) $R = \dfrac{\sqrt{N}}{4}\left(\dfrac{\alpha - 1}{\alpha}\right)\left(\dfrac{k_B}{1 + k_{\text{ave}}}\right)$

$$k_B = \dfrac{7.028\,\text{min} - 1.44\,\text{min}}{1.44\,\text{min}} = 3.88$$

$$\text{Calculate } k \text{ values:}\quad k_A = \dfrac{6.912\,\text{min} - 1.44\,\text{min}}{1.44\,\text{min}} = 3.80$$

$$k_{\text{ave}} = \dfrac{3.88 + 3.80}{2} = 3.84$$

$$\text{Calculate } \alpha:\quad \alpha = \dfrac{k_B}{k_A} = \dfrac{3.88}{3.80} = 1.021$$

$$R = \dfrac{\sqrt{N}}{4}\left(\dfrac{\alpha - 1}{\alpha}\right)\left(\dfrac{k_B}{1 + k_{\text{ave}}}\right) = 1.196$$

$$1.196 = \dfrac{\sqrt{N}}{4}\left(\dfrac{1.021 - 1}{1.021}\right)\left(\dfrac{3.88}{1 + 3.84}\right)$$

$$1.196(4)\left(\dfrac{1.021}{0.021}\right)\left(\dfrac{4.84}{3.88}\right) = \sqrt{N}$$

$$290 = \sqrt{N}$$

$$84{,}100 = N = \text{plates}$$

(c) $H = L/N = 10.0\,\text{cm}/84{,}100\,\text{plates} = 1.19 \times 10^{-4}\,\text{cm/plate}$ (really quite good for an RPLC column).

**1.14** (a) Column 2 produced the broader peak.

(b)

$$N_1 = 5.54\left(\dfrac{10.20\,\text{min}}{0.072\,\text{min}}\right)^2 = 111{,}000\,\text{plates}$$

$$N_2 = 5.54\left(\dfrac{24.965\,\text{min}}{0.138\,\text{min}}\right)^2 = 181{,}000\,\text{plates}$$

Column 2 has more theoretical plates.

(c) Column 2 is better in terms of total plates.

(d) Peak width alone is not an indicator of the number of theoretical plates, and therefore of the resolving power of a column. The time of elution must also be considered. Column 2 produced the broader peak but ultimately has more theoretical plates because the broadening *per unit time* is better on column 2 than on column 1.

**1.15**  (a) Note that the problem focuses entirely on *longitudinal* diffusion. (a) Diffusion coefficients in gas are 10,000–100,000 times greater than diffusion coefficients in liquids. So the approximation is good because the longitudinal spreading that occurs in the stationary phase will be quite small due to the slow diffusion in a polymer film. The longitudinal diffusion that occurs in the gas phase, however, is significant because of the much higher diffusion coefficients in gas. So it is appropriate to say that the longitudinal broadening in the stationary phase is negligible compared to that which occurs in the mobile phase.

(b) The approximation is not equally valid when the mobile phase is a liquid because now the diffusion coefficients in the mobile and stationary phases are comparable – so longitudinal diffusion that occurs in the stationary phase could potentially be comparable to that which occurs in the mobile phase.

**1.16**  As temperature increases, the viscosities of the mobile and stationary phases decrease. This, in turn, makes it easier for solute molecules to diffuse in them, increasing the diffusion coefficients, $D_s$ and $D_m$. The $C$-terms depend on the inverse of these diffusion coefficients, such that an increase in diffusion coefficients leads to a decrease in plate heights (i.e., more favorable $H$).

## GAS CHROMATOGRAPHY

**2.1**

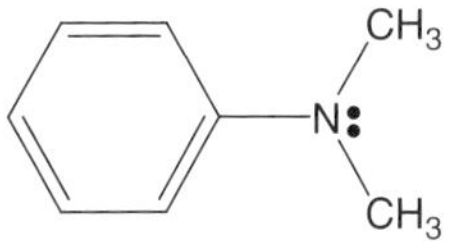

*N,N*-Dimethylaniline

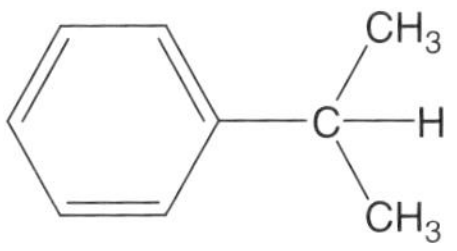

Cumene

*N,N*-Dimethylaniline is more polar and should therefore be more retained on the polar Carbowax column.

**2.2**

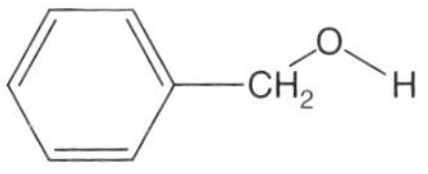

Benzylalcohol

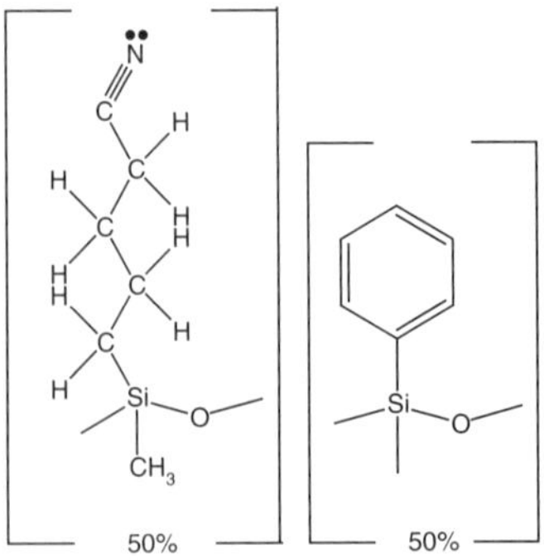

Interactions:

- Hydrogen bonding (benzyl alcohol donating the hydrogen to the cyano group)
- Dipole–dipole
- Dipole–induced dipole
- Dispersion

**2.3**

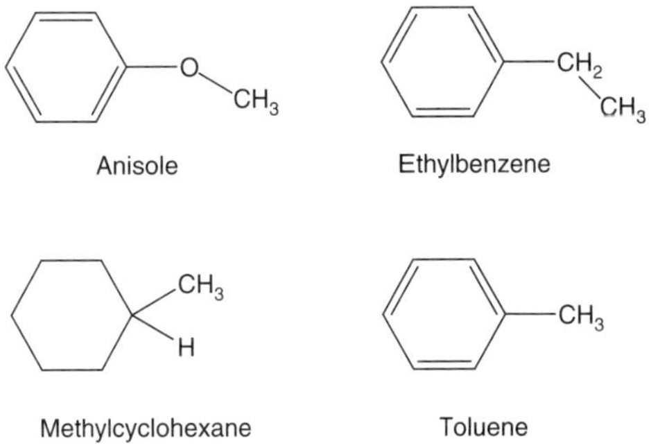

Anisole

Ethylbenzene

Methylcyclohexane

Toluene

(a) Moderate polarity columns will retain anisole slightly more than ethylbenzene.

(b) Columns that maximize dispersion and dipole–induced dipole interactions will retain toluene more than methylcyclohexane due to the enhanced polarizability of the aromatic ring. Thus, columns with phenyl rings or polar columns would maximize the separation.

In truth, both sets of compounds could likely be separated in most GC columns simply through judicious temperature selection. Nevertheless, it is useful to think about the intermolecular interactions that would maximize the separation.

**2.4**

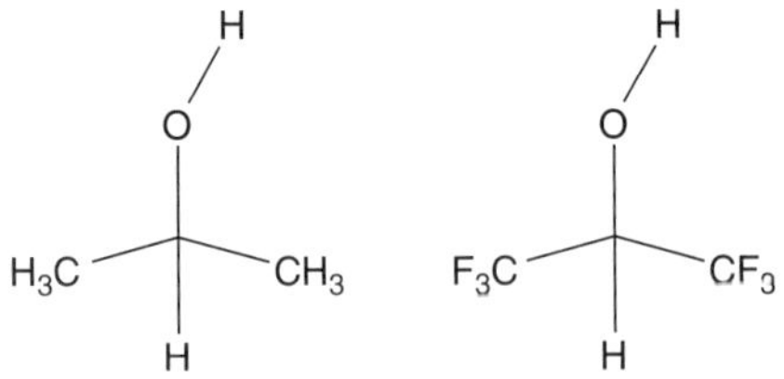

(a) HFIPA is the stronger hydrogen bond donor due to the electron-withdrawing effect of fluorine, which draws electron density away from the O—H bond, thereby creating a larger partial positive charge on the hydrogen atom, which increases its strength of hydrogen bonding.

(b)

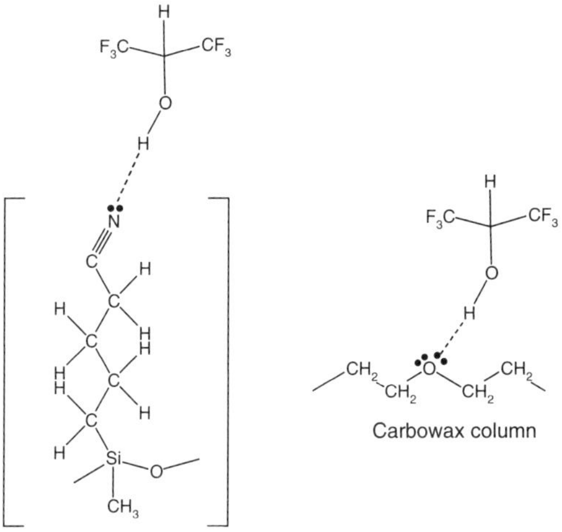

Cyanopropyl phase

Carbowax column

(c) $\alpha = \dfrac{k_{\text{HFIPA}}}{k_{\text{IPA}}} = \dfrac{0.589}{0.186} = 3.17$

(d) The separation factor will decrease. As temperature increases, all intermolecular interactions decrease in strength (to a limit of no interactions at high temperature). Thus, as temperature increases, the influence of the difference in hydrogen-bonding strength between the two compounds will decrease in importance and they will elute closer to one another.

**2.5** (a)

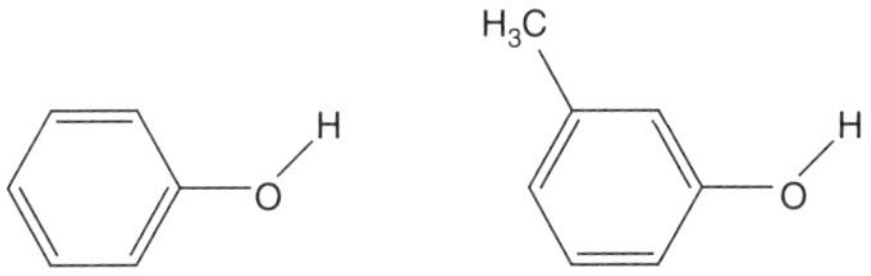

Phenol

m-Cresol

(b) m-Cresol will elute after phenol on all columns due to its increased dispersion interactions arising from the additional methyl group. All other intermolecular interaction abilities of the two compounds are nearly identical.

(c) DB-1 phases are nonpolar and retain phenol via dispersion and dipole–induced dipole forces. DB-225 and wax phases are polar phases and are capable of accepting hydrogen bonds. Thus, they retain phenol via hydrogen bonding and dipole–dipole interactions in addition to dispersion and dipole–induced dipole forces.

**2.6** (a) $\Delta H_{\text{vap, hexane}} = 31\,\text{kJ/mol}$
$\Delta H_{\text{vap, heptane}} = 36\,\text{kJ/mol}$
$\Delta H_{\text{vap, methanol}} = 37\,\text{kJ/mol}$

$\Delta H_{vap}$ is a measure of the ease of getting bulk phase liquid molecules into the gas phase. Thus, the predicted GC elution order would be as follows: hexane, heptane, and methanol.

(b) Methanol elutes first rather than last as predicted using vaporization enthalpies. This occurs because dipole–dipole and hydrogen-bonding forces exist between molecules in pure liquid methanol. These forces increase the energy required for vaporization. Inside the column, however, the methanol molecules do not interact with one another but rather with the PDMS stationary phase. This phase cannot interact via dipole–dipole and hydrogen-bonding forces. Thus, the intermolecular interactions experienced by the molecules inside the column are weaker than those that contribute to methanol's relatively high enthalpy of vaporization. Because the interactions are weaker with the stationary phase than with other methanol molecules, methanol elutes earlier than predicted based on its bulk phase $\Delta H_{vap}$.

Conversely, hexane and heptane experience only dispersion interactions in the bulk phase and only dispersion interaction with the PDMS phase. Because the intermolecular interactions are of the same type and strength in both environments, these compounds elute in the order predicted by their vaporization enthalpies (i.e., hexane before heptane).

(c)

$$\alpha_{hexane/methanol} = \frac{0.161}{0.0711} = 2.26$$

$$\alpha_{heptane/hexane} = \frac{0.343}{0.161} = 2.13$$

(d) When bulk phase interactions are of the same type and strength as those experienced in the column, boiling points and enthalpies of vaporization can lead to useful predictions of relative retention. When they are not similar, differences between the predicted and observed elution orders may be observed.

**2.7** (a) Decreased

(b) Decreased

**2.8** Derivatization can make a molecule more volatile by converting a polar functional group into one that is less polar. In addition, derivatization is sometimes used to "tag" a molecule with atoms or functional groups that are specific for a particular detector to enhance the sensitivity and selectivity of the analysis.

**2.9** (a) Enzymes and macromolecules are not volatile enough to be analyzed by GC. To elute from the column, solutes must exist in the gas phase for a period of time. Enzymes and macromolecules are so large that their intermolecular interactions with the stationary phase would be so great that they would never get into the gas phase. In addition, if temperatures are increased too high to try to vaporize them, they may degrade. Furthermore, there is a limit (typically 300–400 °C) to temperatures that can be applied to GC columns.

(b) Similarly, ions are not volatile and cannot be analyzed by GC.

**2.10** The carrier gas sweeps the solutes down the column when they are in the gas phase. It does not interact with the solutes. Solute peaks are broadened both at high and low flow rates (see the van Deemter equation in Chapter 1). Broader peaks have a better chance of overlapping and thereby degrading resolution. Furthermore, at very high flow rates, all of the solutes would be swept to the end of the column so quickly that they'd have no chance to interact with the stationary phase. This clearly would cause resolution problems. Typical flow rates for conventional GC with capillary columns are between 1 and 5 mL/min.

**2.11**  (a)  $\log k = \log \left( \dfrac{t_r - t_m}{t_m} \right) = \log \left( \dfrac{0.705 - 0.275}{0.275} \right) = 0.194$   for octane.

Same equations for the others.

| n-Alkane | $t_r$ (min) | $\log k$ |
|----------|-------------|----------|
| Octane | 0.705 | **0.194** |
| Nonane | 1.169 | **0.512** |
| Decane | 2.111 | **0.824** |
| Undecane | **4.045** | **1.138** |
| Dodecane | 8.179 | **1.459** |
| Tridecane | 16.703 | **1.776** |
| Tetradecane | 34.200 | **2.091** |

(b)

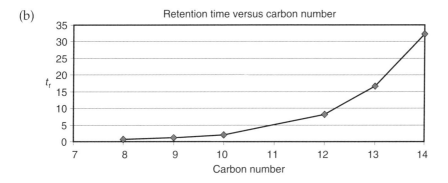

(c)

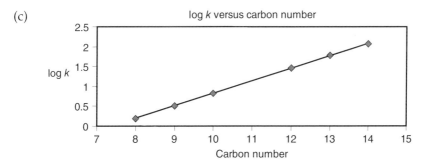

See the discussion that follows Equation 2.21 in the text

$$\Delta G^{\circ}_{CH_2} = -2.303\,RT\,(\text{slope})$$

$$= -2.303(8.314\,\text{J/mol K})(353.15\,\text{K})(0.3162) = -2138\,\text{J/mol}$$

(d) Equation for the best straight line: $\log k = 0.3162 \times CN - 2.34$

$$\log k = (0.3162)(11) - 2.34 = 1.138$$
$$k = 10^{1.138} = 13.74$$
$$k = \frac{t_r - t_m}{t_m} = 13.74 = \frac{t_r - 0.275}{0.275}$$

$t_r = 4.05\,\text{min}$ (value varies depending on rounding within the calculation).

**2.12** (a)

$$\frac{1\,\text{sample}}{15\,\text{min}} \times \frac{60\,\text{min}}{h} \times \frac{24\,h}{\text{day}} \times \frac{7\,\text{days}}{\text{week}} = 672\,\text{samples/week}$$

$$\frac{1\,\text{sample}}{12\,\text{min}} \times \frac{60\,\text{min}}{h} \times \frac{24\,h}{\text{day}} \times \frac{7\,\text{days}}{\text{week}} = 840\,\text{samples/week}$$

(b)

$$\frac{1\,\text{sample}}{16\,\text{min}} \times \frac{60\,\text{min}}{h} \times \frac{24\,h}{\text{day}} \times \frac{7\,\text{days}}{\text{week}} = 630\,\text{samples/week}$$

$$\frac{1\,\text{sample}}{18\,\text{min}} \times \frac{60\,\text{min}}{hr} \times \frac{24\,hr}{\text{day}} \times \frac{7\,\text{days}}{\text{week}} = 560\,\text{samples/week}$$

(c) the isothermal method

(d) 70 sample difference $\times \frac{\$150}{\text{sample}} = \$10,500$

**2.13** Nondestructive methods leave the molecules intact for potential subsequent analysis or collection. Destructive methods degrade the molecule by breaking bonds.

Nondestructive methods: TCD, IR, ECD (only a small percentage of molecules get ionized), PID.

**2.14** (a) Flame photometric detector or perhaps a nitrogen phosphorus detector

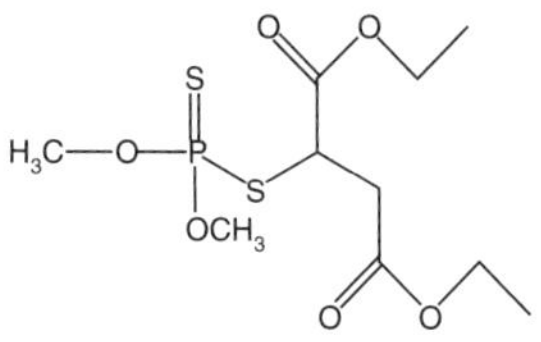

Malathion

(b) Flame ionization detector

$$CH_3(CH_2)_{12}CH_3$$

(c) Flame photometric detector, electron capture detector

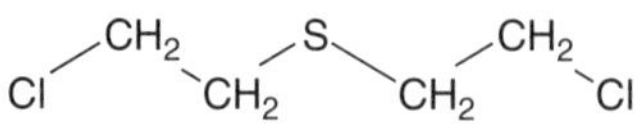

Mustard gas

(d) Flame ionization detector

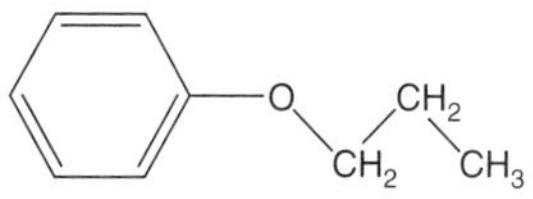

Phenyl propyl ether

(e) Photoionization detector, flame ionization detector

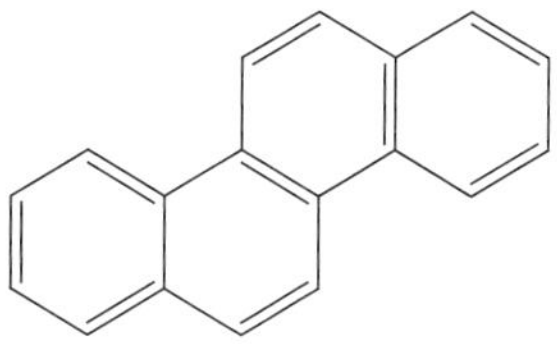

Chrysene

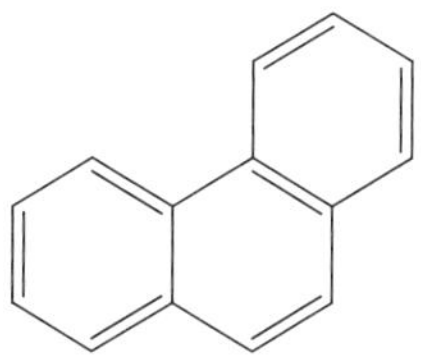

Phenanthrene

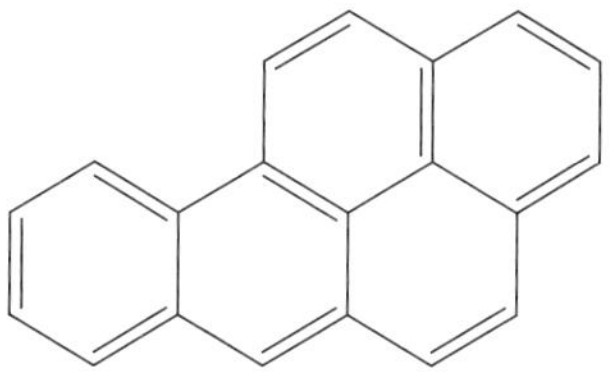

Benzo[a]pyrene

(f) Nitrogen phosphorus detector, photoionization detector

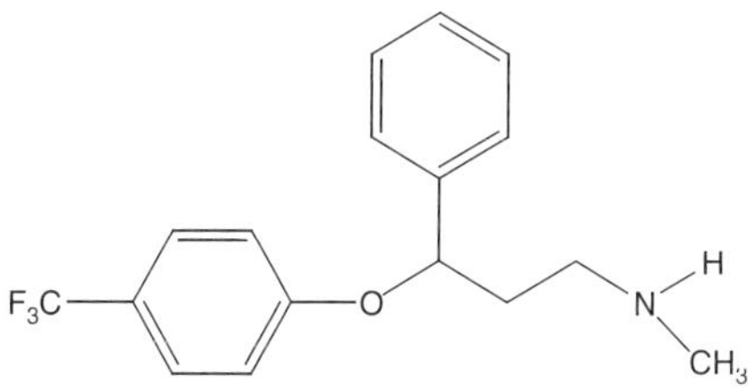

Fluoxetine (Prozac)

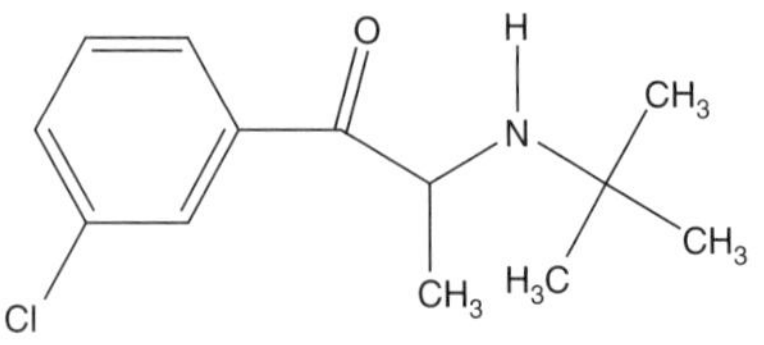

Bupropion (Wellbutrin)

**2.15**  The rare gases are the noble gases: He, Ne, Ar, Kr, Xe, Rn. A packed column or PLOT column would be needed to have enough stationary phase to retain the gases and a TCD might be used to detect them (an FID would not work because the gases do not burn).

**2.16**  (a) Equation for the best straight line: MTBE peak area = 4423 (%MTBE) − 197.4
For peak area = 5289, %MTBE = 1.24.

(b) For the standard deviation in a calculated result from a linear regression:

$$s_X = \frac{s_r}{b} \left\{ \frac{1}{m} + \frac{1}{n} + \frac{(\overline{Y}_X - \overline{y})^2}{b^2 \Sigma (x_i - \overline{x})^2} \right\}^{1/2}$$

| | | |
|---|---|---|
| $s_X$ | = | standard deviation of calculated result |
| $s_r$ | = | standard deviation about the regression |
| $b$ | = | slope of the line |
| $m$ | = | number of replicate analyses used to establish the unknown signal |
| $n$ | = | number of standards used to establish the calibration curve |
| $\overline{Y}_X$ | = | average signal for the unknown |
| $\overline{y}$ | = | average signal for the standards |
| $x_i$ | = | individual concentration (percent) of a standard |
| $\overline{x}$ | = | average concentration of the standards |

*Source:* From Harvey.[1]

For the least squares line in this problem:

$$s_X = \frac{166.2}{4423} \left\{ \frac{1}{3} + \frac{1}{5} + \frac{(5829 - 3784)^2}{(4423)^2 (1.075)} \right\}^{1/2} = 0.032$$

So the %MTBE is $1.240 \pm 0.036$.

(c)

| %MTBE in standard solution | Peak area | | |
|---|---|---|---|
| | MTBE | Butyl ether | Ratio |
| 0.25 | 1025 | 4097 | 0.250 |
| 0.50 | 1975 | 3955 | 0.499 |
| 1.00 | 3997 | 4000 | 0.999 |
| 1.25 | 5365 | 4280 | 1.253 |
| 1.50 | 6557 | 4358 | 1.505 |

Equation for the best straight line: Peak area ratio = 1.004 (%MTBE) − 0.002.

(d) For peak area ratio $= 5289/4235$, %MTBE $= 1.246$.

(e) Same equation as above

$$s_X = \frac{0.00179}{1.004}\left\{\frac{1}{3} + \frac{1}{5} + \frac{(1.246 - 0.9014)^2}{(1.004)^2(1.075)}\right\}^{1/2} = 0.0014$$

So the %MTBE is $1.246 \pm 0.001$.

Note: This problem is designed to illustrate the fact that internal standards lead to improved standard deviations (uncertainties) in concentrations calculated from calibration curves.

The analysis for the ratios likely underestimates the standard deviation because it treats the uncertainty in the ratio the same as just the peak areas. However, because there is uncertainty in each area, taking the ratio of areas would compound the uncertainty in the ratio to some extent. Thus, the mathematical treatment may not be completely rigorous, but it does serve to illustrate that internal standards can effectively remove variability in peak areas (largely caused by the injection process as discussed in the chapter).

**2.17** (a)

$$F^\circ_{corr} = F_m \cdot \frac{T_C}{T} \cdot \frac{P - P_{H_2O}}{P} \cdot j$$

$$j = 1.5\left(\frac{(P_i/P)^2 - 1}{(P_i/P)^3 - 1}\right)$$

$$j = 1.5\left(\frac{(181.8\,\text{kPa}/101.0\,\text{kPa})^2 - 1}{(181.8\,\text{kPa}/101.0\,\text{kPa})^3 - 1}\right) = \left(\frac{1.8^2 - 1}{1.8^3 - 1}\right) = 1.5\left(\frac{2.24}{4.832}\right) = 0.695$$

$$F^\circ_{corr} = (5.20\,\text{mL/min}) \cdot \left(\frac{373.15\,\text{K}}{295.15\,\text{K}}\right)$$

$$\cdot \left(\frac{101.0\,\text{kPa} - 2.64\,\text{kPa}}{101.0\,\text{kPa}}\right) \cdot (0.695)$$

$$F^\circ_{corr} = 4.45\,\text{mL/min}$$

(b)

$$V^\circ_R = t_R \cdot F^\circ_{corr}$$

$$V^\circ_R = (3.73\,\text{min})(4.45\,\text{mL/min}) = 16.60\,\text{mL}$$

$$V^\circ_M = t_M \cdot F^\circ_{corr}$$

$$V^0_M = (0.43\,\text{min})(4.45\,\text{mL/min}) = 1.91\,\text{mL}$$

(c)

$$V_N = V_R{}^\circ - V_M{}^\circ$$

$$V_N = 16.60\,\text{mL} - 1.91\,\text{mL} = 14.69\,\text{mL}$$

**2.18**  Injection discrimination refers to the preferential delivery from a syringe of some solutes over others in a mixture, often arising from volatility differences. Inlet discrimination refers to the preferential delivery of some solutes in a mixture onto the column due to events that occur within the injection port as the solutes are vaporized and swept into the column. Both forms of discrimination introduce a systematic bias in that the quantity of some solutes that make it onto the column are lower than they ought to be.

**2.19**  In split injection, the purge valve is open during the injection. In splitless injection, the purge valve is closed during the injection and opened only after a specified period of time has elapsed. Split injections are used with concentrated samples to avoid column overload, which leads to poor peak shape.

**2.20**  Carryover refers to the introduction of solutes from one sample vial into the next sample vial to be analyzed. It can be minimized by rinsing the syringe with clean solvent between analyses and changing the cleaning solvent frequently.

**2.21**  Some aspects of the injection process are chaotic and random. Solutes are typically delivered in liquid form from a syringe, expand as they vaporize, swirl around in the injection port, potentially make contact with the liner, and are swept into the column or out of the purge vent. Furthermore, small sample volumes are delivered (typically 1 μL or less). It is difficult to reproducibly draw up and deliver such small quantities. All of this makes the injection process potentially imprecise. Internal standards help improve the data by introducing a known quantity of sample that undergoes the same random events and injection uncertainty and thereby help to ratio out some of the variability.

**2.22**  Solid phase microextraction. It is used to preconcentrate solutes from liquid and gas samples and subsequently introduce them into the injection port. The SPME fiber is coated with a polymer film. Solutes partition out of the sample and into the film. The fiber is then introduced into the hot injection port of a GC, where the solutes desorb from the fiber and are swept into the column.

**2.23**  Total area = 258,000 arbitrary units.

$\%n$-Hexane $= 100 \times (67,000/258,000) = 25.97\%$.

| Analyte | Peak area (arbitrary units) | Raw area (%) | Actual (%) |
|---|---|---|---|
| $n$-Hexane | 67,000 | 25.97 | 22.2 |
| $n$-Propyl benzene | 70,000 | 27.13 | 27.1 |
| Naphthalene | 54,000 | 20.93 | 31.3 |
| $t$-Butylnaphthalene | 67,000 | 25.97 | 19.4 |

Taking $n$-hexane as a standard yields $67,000/14$ C—H bonds $= 4785.7$ arbitrary units/C—H group.

Applying this to $n$-propylbenzene,

$$\frac{70,000\,\text{abu}}{} \times \frac{\text{C–H bond}}{4787.7\,\text{abu}} \times \frac{}{12\ \text{C–H bonds}} = 1.22$$

This says that the number of $n$-propylbenzene molecules is 1.22 times that of hexane. For naphthalene,

$$\frac{54{,}000\,\text{abu}}{4787.7\,\text{abu}} \times \frac{\text{C--H bond}}{8\ \text{C--H bonds}} = 1.41$$

For $t$-butylnaphthalene,

$$\frac{67{,}000\ \text{abu}}{4787.7\,\text{abu}} \times \frac{\text{C--H bond}}{16\ \text{C--H bonds}} = 0.875$$

All of the compounds must add up to 100%, so let the amount of hexane be "$x$." Doing so yields

$x + 1.22x + 1.41x + 0.875x = 100\%$

$x = 22.2\% = \text{hexane percent}$

$(1.22)(22.2\%) = 27.1\% = n\text{-propylbenzene percent}$

$(1.41)(22.2\%) = 31.3\% = \text{naphthalene percent}$

$(0.875)(22.2\%) = 19.4\%$

**2.24** (a)

$$RI = 100\left(\frac{\log k_s - \log k_n}{\log k_{n+1} - \log k_n}\right) + 100(n)$$

$$k_s = \frac{t_r - t_m}{t_m} = \frac{333\,\text{s} - 18\,\text{s}}{18\,\text{s}} = 17.5 \quad \log k_s = \log 17.5 = 1.243$$

$$k_n = \frac{t_r - t_m}{t_m} = \frac{194\,\text{s} - 18\,\text{s}}{18\,\text{s}} = 9.78 \quad \log k_s = \log 9.78 = 0.990$$

$$k_{n+1} = \frac{t_r - t_m}{t_m} = \frac{370\,\text{s} - 18\,\text{s}}{18\,\text{s}} = 19.56 \quad \log k_s = \log 19.56 = 1.291$$

$$RI = 100\left(\frac{1.243 - 0.990}{1.291 - 0.990}\right) + 100(10)$$

$$RI = 100(0.8406) + 1000 = 84.06 + 1000$$

$$RI = 1084$$

(b) Decrease. Benzaldehyde is polar and can donate a hydrogen bond to the cyano propylphenyl phase and interact via dipole–dipole interactions. A diphenyldimethylpolysiloxane phase cannot interact via dipole–dipole or hydrogen-bonding interactions. Thus, the retention of benzaldehyde should decrease relative to the alkanes.

(c) Pentane. It is the solvent and thus constitutes the majority of what is injected into the column. For this reason, solvent peaks are usually much larger than analyte peaks.

(d) It is a ghost peak that arises from a solute injected during the previous analysis. It is broader than the other peaks near it because it has spent a considerable amount of time in the column – all of the previous analysis time plus the 240 s during the current analysis – and has therefore had time to diffuse in the axial direction. Lengthening the analysis time and increasing the temperature via temperature programming would have helped elute it during the previous analysis.

(e) Retention times are not thermodynamic properties. Retention factors are. Because $\Delta G° = -RT(\ln k + \ln \beta)$, one expects $\log k$, not retention time, to increase linearly with carbon number. In fact, retention time increases exponentially with carbon number.

(f) Temperature would be the main variable to change. A decrease in temperature would help improve the separation of hexane from pentane. Decreasing the flow rate may also help some. Lastly, if the chromatogram was obtained using direct injection, using slit or splitless injection may help eliminate some of the solvent tail and thereby reduce the overlap.

(g) Temperature programming could be used to help decrease the retention time of undecane and the late-eluting peak that appears as a ghost peak in this chromatogram. Using a temperature program to elute the later peaks faster will not only decrease the analysis time but will also decrease the peak broadening and increase the peak height, which is beneficial for quantitative analysis. A potential drawback to a temperature program would be the possibility that the unknown and undecane might not be as well resolved depending on the severity of the temperature gradient and the temperatures used. In addition, the time it takes for the oven to cool to its starting temperature before each analysis could increase the time per analysis.

**2.25** (a) 100% dimethyl polysiloxane phase (DB-1, Rtx-1)

(b) Nonpolar

(c) Oil is largely comprised of relatively nonpolar hydrocarbons

(d) 60 m column length, 1 μm film thickness, id =0.25 mm

(e) FID – appropriate for hydrocarbon oil analysis because hydrocarbons burn well and therefore generate strong signals in an FID

(f) Split injection

(g) Temperature programmed.

**2.26** By injecting a blank of pure solvent before, during, and after all of the analyses, the analysts are checking the reproducibility of the instrument. If at any point in the data collection the hexane "sample" shows anything other than the peak it showed for hexane in the very first two analyses, they will know there is a problem and will not trust the data. For example, if the retention time of the hexane changes, or if other peaks start to appear in the blank, then something has changed (the flow, column temperature, etc.) that could affect the analyses of the actual samples. It is smart to include this kind of check at the start, finish, and throughout a series of analyses. If no problems are observed and the last hexane chromatogram looks exactly like the first hexane chromatogram, this lends great confidence that the instrument was working consistently throughout the data collection. (Note: While the idea is a good one, it may have even been better to use a mixture that contains some solutes in addition to the hexane – that way, peak shape, area, and height could also be monitored in addition to retention times.) Because solvents overload the column, they generally produce poor peak shapes and off-scale signals. Thus, it is hard to use pure solvents to monitor the consistent performance of the detector.

If all is going well, the analysts should see the exact same chromatogram for each hexane injection interspersed with the actual oil sample chromatograms.

**2.27** See your instructor.

## LIQUID CHROMATOGRAPHY

**3.1** Retention time order: aniline $<$ 4-ethylphenol $< N,N$-diethylaniline $<$ anthracene.

Aniline is small, polar, and can hydrogen bond. Therefore, it will interact with water and organic modifier in the mobile phase the best and elute first. 4-Ethylphenol is also small and polar, but the ethyl group makes it more hydrophobic than aniline and thus slightly more retained. $N,N$-Diethylaniline is structurally related to aniline, but ethyl groups replace the hydrogen atoms on the nitrogen. Therefore, $N,N$-diethylaniline cannot donate hydrogen bonds to water and modifier in the mobile phase. The ethyl groups also make it more hydrophobic. Both of these effects mean that $N,N$-diethylaniline will more favorably interact with the stationary phase relative to the mobile phase compared to aniline and 4-ethylphenol. Finally, anthracene has no ability to donate or accept hydrogen bonds and it is nonpolar. This means it will partition out of the mobile phase and into the stationary phase more readily than the other compounds and thus be most retained.

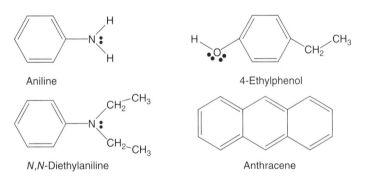

**3.2** (a) Size exclusion chromatography
(b) Ion-exchange chromatography
(c) Affinity chromatography
(d) Reversed-phase liquid chromatography
(e) Reversed-phase liquid chromatography, normal phase liquid chromatography
(f) Use a chiral stationary phase or chiral compounds in the mobile phase.

**3.3** $La^{3+}$, $Al^{3+}$, $Ba^{2+}$, $Pb^{2+}$, $Sr^{2+}$, $Ca^{2+}$, $Ni^{2+}$, $Cu^{2+}$. All of these are more strongly retained and therefore would displace zinc ions.

**3.4** (a) Convert the p$K_b$ values into p$K_a$ values for easier comparison: aniline p$K_a = 4.6$, diethylamine p$K_a = 10.7$.

At pH$= 3$, based largely on extent of protonation, retention times from shortest to longest would be butylsulfate $\approx$ chloroethanoic acid $<$ benzoic acid $<$ aniline $<$ diethylamine.

Rationale: Butylsulfate carries a full negative charge at all pHs and will not interact with the sulfonate group. A pH of 3.8 is a full pH unit above the $pK_a$ of chloroethanoic acid. It will therefore be largely deprotonated and carry a negative charge. Thus, it will not interact with the stationary phase and will elute early with the butylsulfate. Benzoic acid will also be partially charged at this pH but not to the same extent (nearly 50:50), so some retention may result from dispersion interactions but not much. At a pH of 3.8, some aniline will be in the neutral and some in the protonated form. This will increase the attraction to the sulfonate stationary phase and therefore cause greater retention than for neutral compounds. Diethylamine will be completely in its protonated state and will therefore have the most attraction to the stationary phase and be most retained.

At $pH = 8$, based largely on extent of protonation, retention times from shortest to longest would be butylsulfate $\approx$ chloroethanoic acid $\approx$ benzoic acid $<$ aniline $<$ diethylamine.

Rationale: Butylsulfate, chloroethanoic acid, and benzoic acid would all be fully deprotonated and bear a negative charge. Thus, they would essentially coelute. Aniline would be in its fully neutral form and thus may be retained a little via dispersion forces. Diethylaniline, with a $pK_a$ of 10.7, would still be protonated and thus be significantly attracted to the negatively charged sulfonate stationary phase.

(b) Butylsulfate does not protonate in aqueous media (i.e., the conjugate acid is a strong acid).

(c) The lone electrons that give aniline its basicity can be conjugated into the aromatic ring. This dramatically decreases their availability for forming covalent bonds with protons, thus giving rise to the weak basicity of aniline relative to amines in which conjugation effects are absent.

**3.5** No, the mobile phase does not flow through the particles. The pores are too small and resist flow through them. Solutes get into and out of the pores via diffusion. In contrast, in monolithic columns, the "pores" or channels are continuous and allow the mobile phase to transport solutes to the stationary phase in the mesopores.

**3.6** (a) much smaller than (it will not fit into many pores)
(b) smaller than (it may fit into some pores, but not as many as urea does)
(c) approximately equal to
(d) smaller than or much smaller than, depending on the size distribution of the curves.

**3.7** At pH less than 2, the Si—O—Si bonds that link the stationary phase monomers to the silica particle surface begin to hydrolyze. The hydrolyzed monomer then elutes from the column in the mobile phase. Thus, the stationary phase is being systematically removed, which will result in decreasing retention in subsequent analyses.

At pH greater than 8, the silica particles themselves start to dissolve in the mobile phase. This causes the particles to collapse and leads to large voids in the column resulting in poor peak shapes and irreproducible retention times.

**3.8** (a) Change the mobile phase to include more water (more water usually increases separation of organics).

(b) Change the modifier to methanol or acetonitrile. Different modifiers have different interactions with solutes so a different modifier might offer a slightly different selectivity that helps separate the compounds.

(c) Change the column to one with a different stationary phase. Try a phase that is more polar to get improved differential interactions with the polar solutes.

(d) Change the column to one with smaller particles – smaller particles decrease the peak width and may help resolve closely eluting peaks.

The suggestions are listed in order of the cheapest and easiest (a and b) to more time consuming and expensive (c and d). In most LC systems, changing the mobile phase composition is just a matter of specifying the desired composition in the instrument software. New columns, on the other hand, cost several hundred dollars and require instrument downtime to install. The downtime is also a loss of money in many cases.

**3.9**

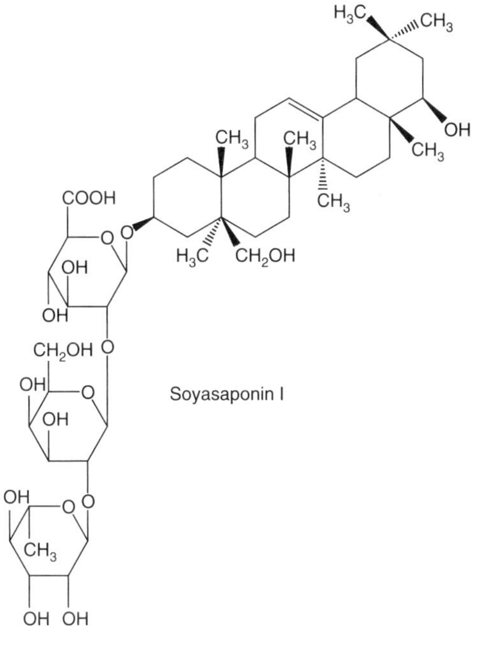

Soyasaponin I

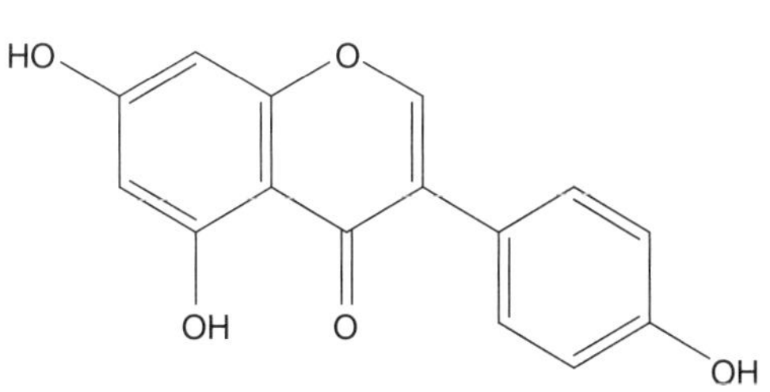

Genistein

Soyasaponin lacks conjugated double bonds and aromaticity – features that genistein possesses. These features, along with lone electron pairs that can be conjugated into the $\pi$-system, help make compounds absorb in the UV–visible region of the spectrum.

**3.10** (a) Conductivity

(b) Refractive index (UV–visible would not work)

(c) UV–visible or fluorescence (fluorescence would be more sensitive and possibly more selective depending on the mixture)

(d) Evaporative light scattering detector (ELSD)

(e) Mass spectrometer

(f) Electrochemical detector and perhaps mass spectrometry (MS)

**3.11** In ion-exchange chromatography (IEX), the mobile phase contains ions. These ions would generate a high background signal. To detect solute ions, then, would require detecting small changes in the conductivity of the solute zone relative to the mobile phase. Ion suppression decreases the conductivity of the mobile phase and therefore allows easier detection of the higher-conducting solute zone.

**3.12**

Chiral carbon

L-Dopa

(a)

D-Dopa

(b)

(c) L-Dopa is biologically active and used to treat Parkinson's disease.

(d) Any of the chiral phases (e.g., Pirkle, cyclodextrin, polysaccharide) could be tried in attempting to separate L- and D-dopa.

**3.13** (a)

$$V = \pi r^2 l$$

$$V = \pi(0.0023\,\text{m})^2(0.15\,\text{m}) = 2.49 \times 10^{-6}\,\text{m}^3 = 2.5\,\text{cm}^3 = 2.5\,\text{mL}$$

(b)

$$V = \frac{4}{3}\pi r^3$$

$$V = \frac{4}{3}\pi(2.5 \times 10^{-6}\,\text{m})^3 = 6.55 \times 10^{-17}\,\text{m}^3$$

(c) $\left( \dfrac{(0.60)(2.49 \times 10^{-6}\,\text{m}^3)}{6.55 \times 10^{-17}\,\text{m}^3} \right) = 2.28 \times 10^{10}$ particles

**3.14** Endcapping is the use of short alkyl silanes to react with silanol sites on the surface of particles that were not reacted with the stationary phase monomers. It is done to block or modify these sites, removing the silanol that can be very retentive, especially toward bases. Silanols on the surface can also lead to poor peak shapes.

**3.15**   Adopting this scale:

$$5\,\mu m = 5 \times 10^{-6}\,m = 560\,ft$$

$$so\ 1\,m = 1.12 \times 10^{8}\,ft\ on\ this\ scale$$

Column length:

$$\frac{15.0\,cm}{} \times \frac{m}{100\,cm} \times \frac{1.12 \times 10^{8}\,ft}{m} = 1.68 \times 10^{7}\,ft$$

$$\frac{1.68 \times 10^{7}\,ft}{} \times \frac{1\,mile}{5280\,ft} = 3183\,miles$$

Column diameter:

$$4.6\,mm \times \frac{m}{10^{3}\,mm} \times \frac{1.12 \times 10^{8}\,ft}{m} \times \frac{mile}{5280\,ft} = 97.6\,miles$$

So if the solute is the size of a pea, then the column length is comparable to, and in fact a bit longer than, the distance from New York to Los Angeles, and its diameter is nearly 100 miles wide.

**3.16**   Light can cause degradation of organic components. Storing solutions in amber bottles reduces this possibility, and keeping them in refrigerated conditions slows the reaction kinetics of any reactions that might occur. Quantitative results would be artificially lowered if degradation occurs.

**3.17**   A C-13 labeled sulfonamide is a good choice for an internal standard because ideally, internal standards are as chemically close in structure and behavior to the solute as possible. Because isotopes are nearly chemically identical to one another, differing only in their number of neutrons, they satisfy the criteria for good internal standards.

**3.18**   Output of the standard column:

(a)          $$\frac{365\,days}{year} \times \frac{24\,h}{day} \times \frac{60\,min}{h} \times \frac{1.5\,mL}{min} = 788{,}400\,mL/year$$

Output of the narrow bore column:

$$\frac{365\,days}{year} \times \frac{24\,h}{day} \times \frac{60\,min}{h} \times \frac{400\,\mu L}{min} \times \frac{mL}{1000\,\mu L} = 210{,}240\,mL/year$$

The difference in output is therefore 578,160 mL of waste.

$$\frac{578{,}160\,mL}{} \times \frac{1\,L}{1000\,mL} \times \frac{\$30}{L} = \$17{,}345\,saved$$

(b)
$$1\,\text{yr} \times \frac{525,600\,\text{min}}{\text{yr}} \times \frac{1\,\text{sample}}{30\,\text{min}} = \frac{17,520\,\text{samples}}{\text{yr}} \quad \text{for standard RPLC}$$

$$1\,\text{yr} \times \frac{525,600\,\text{min}}{\text{yr}} \times \frac{1\,\text{sample}}{5\,\text{min}} = \frac{105,120\,\text{samples}}{\text{yr}} \quad \text{for fast HPLC}$$

(c) Narrow bore columns can provide more analyses with less waste than standard RPLC columns.

## REFERENCE

1. Harvey, D. *Modern Analytical Chemistry*; McGraw-Hill Science: Dubuque, IA, 1999.

# INDEX

*Chromatography: Principles and Instrumentation*, First Edition. Mark F. Vitha.
© 2017 John Wiley & Sons, Inc. Published 2017 by John Wiley & Sons, Inc.